SOTHIS PRESS

Sothis Press aims to completely reimagine the public domain book with an emphasis on universal access to captivatingly designed print editions. We believe that technology will enable every single book ever published to be perpetually accessible in beautiful print form. Sothis is beauty in ubiquity.

www.sothis.press
1 Waterhouse Square, London, ECIN 2ST

Forging of Iron and Steel

A Text Book for the Use of Students in Colleges, Secondary Schools and the Shop

William Allyn Richards

SOTHIS PRESS

ISBN 978-1-0365-1879-0

Forging of Iron and Steel
is in the public domain.

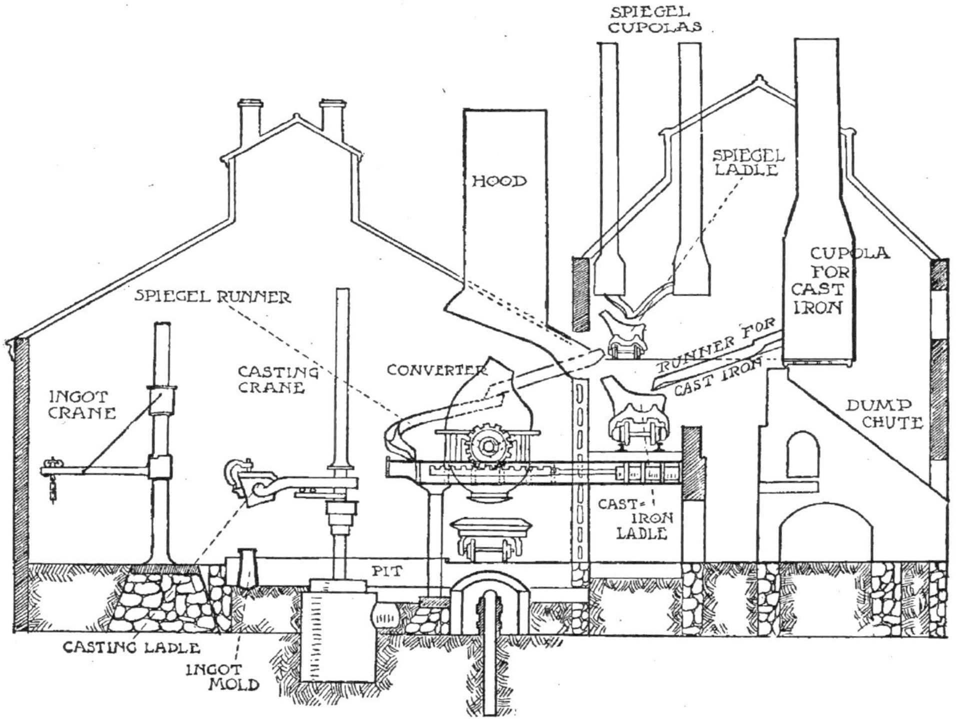

Arrangement of a Bessemer Steel Plant; View showing one Converter Only.

FORGING

OF

IRON AND STEEL

A TEXT BOOK FOR THE USE OF STUDENTS IN COLLEGES, SECONDARY SCHOOLS AND THE SHOP

BY

WILLIAM ALLYN RICHARDS, B.S. IN M.E.

PRINCIPAL OF THE GRANT VOCATIONAL HIGH SCHOOL
CEDAR RAPIDS, IOWA

FORMERLY SUPERVISOR, MANUAL TRAINING SCHOOL, ROCKFORD, ILL.
AND INSTRUCTOR IN FORGE, FOUNDRY, AND MACHINE PRACTICE
IN THE UNIVERSITY HIGH SCHOOL AND UNIVERSITY
OF CHICAGO, CHICAGO, ILLINOIS

337 ILLUSTRATIONS

NEW YORK
D. VAN NOSTRAND COMPANY
25 PARK PLACE
1915

PREFACE

In the preparation of this book, the author has endeavored to treat the forging of iron and steel, and the hardening and tempering of tool steel, simply enough for the High School boy and at the same time thoroughly and systematically enough for the veteran smith. A chapter has been introduced on the history of forging since it is thought to be of interest to all engaged in the work of forging metals. Another chapter on the manufacture of Iron and Steel has been inserted because it is believed that the workman should have some knowledge of the metals, and how they are obtained. It is not thought necessary or advisable to go deeply into the subject of metallurgy, or to introduce metallurgical theory.

No attempt to treat specific exercises has been made; the aim has been to bring out principles. All the methods used toward this end have been thoroughly tried out during ten years of experience in teaching and supervising Manual Training. Those wishing a course of study by which to work, or to use in outlining such a plan, will find in the Appendix a course of study which the author has tried out successfully with several hundred pupils in high school, and with students in college.

In order to obtain the best information possible, the author has consulted nearly every book published on the subject. He has found much of value in the following: *The American Steel Worker*, by E. R. Markham; *Practical Blacksmithing*, by M. T. Richardson; *A Text Book of Elementary Metallurgy*, by Arthur H. Heorns; *Metallurgy of Iron and Steel*, by Bradley Stoughton; *Notes on Iron, Steel and Alloys*, by Forrest R. Jones; *A Handbook of Art Smithing*, by Franz Sales Meyer; *The Smithy and Forge*, by W. J. E. Crane; *Smith's Work*, by Paul N. Hasluck; *Forging*, by John Lord Bacon. Acknowledgment is here made of the use of these and other publications.

W. A. R.

February 15, 1915

CONTENTS

CHAPTER V

CHAPTER VI

CHAPTER VII

CHAPTER VIII

CHAPTER IX

CHAPTER X

CHAPTER XI

CHAPTER XII

CHAPTER XIII

CHAPTER XIV

CHAPTER XV

CHAPTER XVI

FORGING OF IRON AND STEEL

INTRODUCTION

Forging is the process of shaping hot iron or steel by means of a hand hammer, a power hammer, or a press.

This process of shaping may involve any one or all of the following operations:

1. Drawing out.
2. Upsetting.
3. Shaping.
4. Bending.
5. Punching, cutting and splitting.
6. Welding.
7. Hardening and tempering of steel.
8. Annealing steel.
9. Brazing.

1. Drawing out consists of lengthening metal by blows from a hammer, by rolling it between rolls, or by pressing it in a press usually operated by hydraulic power. The shape of the cross section of the metal may or may not be changed in the process. A square section may be made round, a round section made square or hexagonal.

2. Upsetting is the reverse of drawing out. It consists of shortening the length of the piece of metal and increasing the cross section by use of the hammer or press. What was said of the shape of the cross section under drawing out remains equally true in upsetting.

3. **Shaping** or changing the cross section of the piece of metal is also accomplished by means of either the hammer or press. This operation usually combines the first two.

4. **Bending** is done with the tools already mentioned. It is performed on any shaped piece of metal. Usually one side of the piece is stretched, while the other is compressed or upset.

5. **Punching,** splitting and cutting are operations very similar to one another. Punching is making holes of any shape. They usually are round, square, or elliptical, and are made by driving a punch of the proper size and section through the metal by means of blows or pressure. Splitting and cutting are accomplished by driving a chisel through the metal; splitting is usually lengthwise of the piece, and cutting, crosswise, to sever the stock.

6. **Welding** is the uniting by force of two or more pieces of metal, or the two ends of a single piece (bent so as to meet), while heated to such a high temperature that they are plastic, thus allowing their fibers to be joined together. This requires a nice and contemporaneous adjustment of the heat in the parts to be welded. The surfaces must be clean.

7. **Hardening** and tempering are performed on tool or crucible steel. Hardening is accomplished by the sudden cooling of steel that has been heated to a very definite temperature.[1] The degree of hardness depends upon the chemical content of the steel and the rapidity with which it is cooled. **Tempering** is the slightly softening or toughening of a piece of hardened steel, by the

[1] Steel will harden at all temperatures above that of a very dull red; but there is one definite temperature that is correct. This correct temperature and its variation with different steels will be taken up in the text.

process of again heating it to some certain temperature — usually determined by the color of the oxide the heat produces, — and cooling it to prevent further softening.

8. **Annealing** is the softening of a piece of hardened steel by heating it to a definite temperature and then allowing it to cool slowly so that it can be worked with cutting tools or by other means.

9. **Brazing** is the joining together of two or more pieces of metal by means of a brass spelter or one of silver.

All of these operations require special heating, which processes must be learned. The work must be done rapidly while the metal is hot, and the operations must be stopped before the temperature has fallen too low. These matters, as well as how to hold the metal and the hammer, how to strike — whether lightly or heavily, rapidly or deliberately — and what particular tool to use, will be discussed in the chapters which follow.

CHAPTER I

HISTORIC USE OF IRON AND STEEL

THE working of iron and steel is unquestionably one of the oldest of arts. It is known that iron was produced and used at a very early date, probably in prehistoric times.

Early Uses:

In the Bible (Gen. iv, 22) we read of Tubal-cain, son of Lamech and Zillah, as "an instructor of every artificer in brass and iron."

We can properly call Tubal-cain and these early metal workers *smiths*, even though they sometimes worked in brass, as their principal work was the making of armor.

We find abundant references to show that this earliest of trades was held to be highly important. In I Samuel xiii, 19, we find, "Now there was no *smith* found throughout all the land of Israel: for the Philistines said, Lest the Hebrews make them swords or spears;" and we find that Nebuchadnezzar followed the same course among the Jews (II Kings, xxiv, 14), "And he carried away all Jerusalem, and all the princes, and all the craftsmen and *smiths;* none remained, save the poorest sort of the people of the land." Jeremiah xxiv, 1, also states that Nebuchadnezzar carried away the carpenters and *smiths*.

The extent to which the work of these early smiths was carried can be seen from the following references to the Old Testament: Axes, Deut. xix, 5; II Kings vi, 5; stonecutters' tools, Deut. xxvii, 5; armor, coats of mail and weapons of war, I Samuel xvii, 7–38; iron bedsteads, Deut. iii, 11; iron pens, Ezek. iv, 3. We know little about the early smiths and their work, or their method of working the iron. An Egyptian wall painting (Fig. 1), probably gives as reliable an idea as can be found.

The fire was built in a slightly depressed place in the ground; a forced draft was given to the fires by an at-

Fig. 1

tendant on either side, who worked bellows which were placed on the ground in such a manner that they would blow the fire. The attendants worked these bellows by standing on them, alternately throwing their weight from one foot to the other and pulling up the bellows with a rope as the weight was relieved, thus permitting the instruments to be emptied and filled alternately. The little *figure* opposite the smith's head was probably for fuel or water.

The use made of iron by the early Egyptians is a debatable point, for some writers maintain that the Egyptians must have used it in war; while some claim that it was used merely as a precious metal. Egyptian specimens found in tombs and elsewhere are so few that the proof is slight. Recently discovered Egyptian iron finger rings and other articles of personal adornment imply that this metal was scarce and of great value. On the other hand, some of the articles unearthed are an iron-bladed adze with an ivory handle, a thin fragment of wrought iron plate found in an air passage of the Great Pyramid, and an iron blade of a falchion discovered under a Sphinx at Karnac. These articles imply that the metal was somewhat abundant.

Herodotus, the Greek historian, thinks that iron was used generally by the Egyptians for weapons as early as the seventh century B.C. He believes this because when the Carians and Ionians invaded Egypt they were armed with brass and bronze weapons, and an Egyptian, who had never seen arms made of these alloys, ran to inform the king, Psammetichus, of the matter. In Egypt, very few iron weapons have been found, however, whereas many of brass and of bronze have been unearthed. This may be explained, in part, by the fact that iron rusts more than brass or bronze and by the supposition that the brass and bronze weapons belonged to the invading armies.

Iron was known in Assyria and Babylon also. Excavations have brought to light various articles, such as weapons, finger rings, bracelets, chains, hammers, knives and saws. An iron store, the contents of which weighed approximately 385 tons, identified as unwrought ingots, was found at Korsbad. These ingots are pointed at both ends, with a hole near one end, probably so that they might be strung together and more easily transported.

Iron came into use early in Palestine and Phœnicia, also in China, Japan, Persia, and India. It is claimed by the Chinese that steel was invented about 2000 B.C. and that the Indian steel was known fully as early.

The early Greeks and Romans were acquainted with iron, and with them, as with the Egyptians, the first iron probably was of meteoric origin. That bronze was used before iron is recognized by Greek and Roman writers. Hesiod and Homer, Greek poets, have written of bronze, iron and steel. Iron objects have been disinterred at Troy and Mycene. The welding and soldering of iron is said to have been invented by Glaucos of Chios, about 600 B.C. Not only weapons of war were made of iron and steel but also crude farm implements. Iron was used also for ornamental vases and statues. At Delphi, a vessel of silver with a fancifully wrought iron base is described. Hercules is said to have had a helmet of steel and a sword of iron; and Saturn, a steel reaping hook. Diamachus wrote in the fourth century that different kinds of steel were then produced in various places. The best came from Chalybes and India, although steel from Lydia and Laconia was noted. Anvils, pincers, hammers and even the bellows pictured on Grecian vases are similar to those used now.

The early history of the Romans tells us that they were familiar with this metal. Many iron articles have been found in Etruscan and in Roman graves at Pompeii, Vulci, and other places. In many instances, the utensils, weapons, and articles for use were of iron, while those for ornamentation were brass or bronze. It is probable that the early Romans obtained most of their iron from the island of Elba, but after acquiring the sovereignty of the world, they probably mined and manufactured iron in their provinces of Carinthia, Spain, and on the Rhine, where it is presumed that they found well-established industries. In brief, the Greeks and Romans knew iron and its use. They produced it in open hearths or ovens with the assistance of a natural wind draft or bellows, which sometimes produced a material similar to wrought iron and sometimes, steel.

Architectural and Domestic Uses. — During about the tenth century, iron and smiths' work began to be put to architectural and domestic uses. As a rule, the hammer and anvil were the only tools used in producing the artistic specimens which have been handed down to us. We must also consider that the smith of this time had not rolled materials of every form and size which are now obtainable, but that each rod, wire or sheet had to be wrought by himself. We must admit that these modern conveniences have not added to the artistic nature of the product of the smith, but rather have taken away from it. Not only did manual labor produce a better iron than do mechanical operations of the present time, but the outward appearances were more original and interesting than those of machine production, although the latter is without question more exact and neat in appearance.

As machinery came into use, hours of labor shortened, and products cheapened, with the result that large objects could be produced as well as small ones. In early times the smith was compelled to confine most of his labors to small articles, but even when large pieces were undertaken the results were very remarkable.

During the twelfth and thirteenth centuries the work produced by the smith for architectural purposes obtained great importance. Herein the church also became interested and called for ornaments for doors and gateways, window fastenings, chests, and hanging candelabra. Hearth furniture, fire-dogs, wall anchors, and door-knockers were used in castles and other buildings.

Romanesque Period. — The smiths' work of the Romanesque period presents very little beauty in external appearance, but by full and heavy forms it gives the impression of great stability. They followed the simple style of the architecture and the ornaments of the time.

The richest work of this period was produced just before the transition to the Gothic period of Architecture, and is found in door fittings in which case the iron is spread over large flat surfaces. This was probably at first to hold the narrow boards together, but later to furnish the decoration. Thus we find the simple hinge of the first part of the period succeeded by scrolls spreading all over the door. Some characteristics of the Romanesque iron-work are the slit bars and the scrolled parts (Fig. 2), separate bars welded into complex ones and ornaments forged in swages. This work was all forged out of one whole piece of metal or if separate parts were forged they were welded together, so that screws and rivets were not used to hold the parts together. Bonds or ties, however, were used to some extent (Fig. 3).

Fig. 2

Fig. 3

Gothic Period.—Many changes were noticed in the Gothic period. Instead of a forging out of one piece, or pieces forged and welded together, the work consisted of many separate forgings which were riveted to the principal parts. There was made a change in the leaves as the bars were flattened to thin sheets at the ends and cut to shape. Bending, stamping or embossing were also introduced, as were twisted bars. Several tools, such as graving tools, punches and chisels, were added to those already in use. The productions became richer and more elaborate, until Gothic art reached its height, when very elaborate articles were made, such as chandeliers, lanterns and iron furniture. Fancy keys (Fig. 4) were made and hung on a background of colored cloth or leather to bring out the effects. The credit of having made the first attempts to beautify iron with paint and at the same time to preserve it against rust is given to the middle ages.

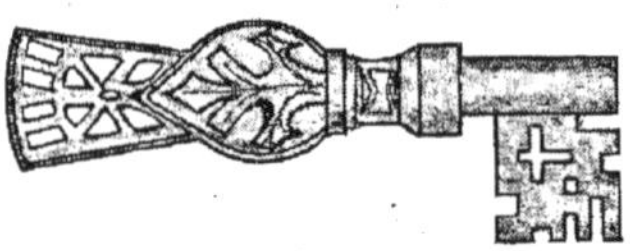

Fig. 4

Renaissance Period.— Following the Gothic period is that of the Renaissance, although Gothic details were not rare up to the fifteenth century or even beyond it. In the southern countries of Europe the wrought ironwork of this period was very simple in appearance, the ornamentation being flat and produced by punching (Fig. 5), but in the Northern countries greater richness was developed. During this period, the use of wrought iron greatly increased and many new articles were introduced, such as reading-desks, wash-stands, towel holders, door grills, brackets, signs, weathercocks and utensils of the most varied kinds.

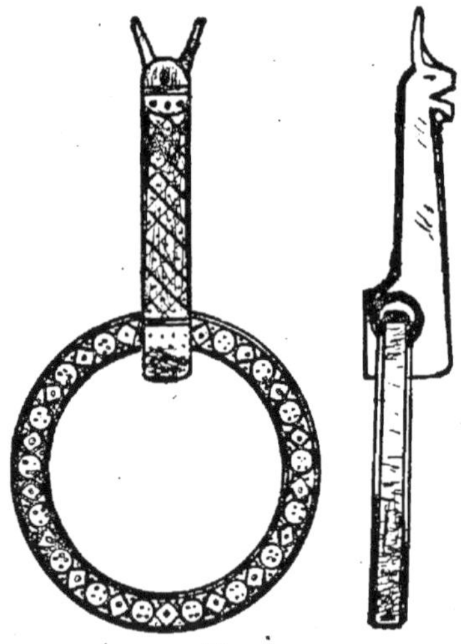

Fig. 5

It was during the renaissance that the production of weapons reached its perfection. Cast iron was also introduced, although it was limited almost entirely to fire backs and stove plate.

The master armorers of Augsburg, Nuremberg and Munich attained great fame at this time. The most costly suits of armor in the museums of Paris, Madrid and Vienna came from the forges of these cities. The designs were furnished by Schwarz, Hirschvogel, Miehich, Floetner, Aldegrever, Dürer, Wohlgemuth and Holbein, the most distinguished artists of the time. These masters added engraving and etching to the earlier arts of embossing and encrusting armor with precious metals. Besides armor they made exquisite shields and sword-hilts, also domestic utensils, tools, instruments of torture, strong boxes, statuettes carved from the solid, and even a throne which was presented to Rudolph II by the Augsburgers in 1574.

German Work. — German iron-work is particularly worthy of study, not only because it is beautiful, but also because it enjoyed a boundless prosperity without a break, from the thirteenth century until the invasion of the first Napoleon, except during the Thirty Years' War. This is exceptional because in Spain, France, England, Italy and the Low Countries, the working of iron ebbed and flowed according to the prosperity of the countries. Blacksmithing was practised from the Rhine to the limits of Austria and from Denmark to Italy, and the greatest variety of articles was produced. Little is known of the work of the early Teutons. The iron hinges and guards on the Romanesque doors, which have withstood time, show a striking resemblance to the work of central France, while others are patterned after the more carefully designed swagework of Paris.

Not until the thirteenth century did the German blacksmiths show independent design. At this time in Marburg, Magdeburg and other places they began using elegant, branching strapwork ending in peculiar *fleur de lis* and vine leaves, which work was a German modification of the French design. This breaking away from the French designs progressed during the next two centuries. The results were a distinct German design in which always appeared the vine, tracery, and *fleur de lis.* In Cologne, on the eve of the Renaissance, about two hundred years later, a new type of work appeared with the thistle as a base for the design (Fig. 6), originated by a family of smiths named Matsys, which produced the celebrated Antwerp well cover. The thistle combined with tracery was in vogue until displaced by Renaissance ornament.

Fig. 6

Baroque Period. — In the "baroque" period the striving after pomp and grandeur produced some very large pieces of elaborate work used mostly in the service of courts and princes. The term, baroque, is understood to mean oval. In this particular style the spirals are squeezed together so as to form ovals, and also are ornamented with foliage (Fig. 7). As this style was used by the architects almost exclusively in the buildings of the Society of Jesus, it is often called the Jesuit style.

Some of the changes from earlier periods are: Round bars gave place to square ones; bars that heretofore were threaded through each other were changed to halving and oversetting; forgings were placed on sheet-iron backings (Fig. 8); leaves became bolder, and rosettes and acanthus husks were used in profusion. The styles for the large pieces were less uniform in appearance than during the Renaissance and earlier periods, since the efforts were spent on the prominent parts, and the less

Fig. 7

important places were left almost bare or filled with straight bars. The treatment of small articles was similar though less striking. The art of casting was becoming better known; therefore parts that were formerly made of wrought iron were made of other materials, and we find the smith's work diminishing.

Rococo Period. — In the eighteenth century came a style known as the rococo. The word "rococo" is derived from "rocaille," which means grotto or shell work. The

Fig. 8

work of this period was very dainty and artistic. The strong heavy grilled windows of the earlier periods gradually became less numerous, probably due to the fact that since the times were less dangerous than before, their use was unnecessary. Balustrades and balcony railings put in an appearance, as well as large iron gates for churches, palaces and parks. The demand for sign brackets and signs for guilds increased. Wrought iron gained more popularity in this line than ever before. The disappearance of straight lines is a particular feature of this period; they were used only when it was absolutely necessary. In place of these we find wild scrollwork. The acanthus similar to that once used in the Gothic was again brought out but the foliage was more crinkled.

Fig. 9

It was evident that the idea was to avoid flat surfaces and to put life into the work in a simple way. Festoons, sprays and garlands were put in every vacant or empty space (Fig. 9).

Deterioration of Art Iron-work. — At the beginning of the reign of Louis XIV in France, the rococo designs had reached their height and the demand for artistic forgework took a turn backward to the more simple style. Grills were made of antique scrolls with interwoven and flowery borders of small stiff design. Wreaths with many bows and ribbons were placed in elliptically shaped shields. This degeneration of art forgework continued down to the first of the nineteenth century. Then for three-quarters of a century little attention was given to it. But, twenty or thirty years ago, in Germany and some other countries, considerable interest sprang up and work is now produced equal to that of any of the old smiths. It is characteristic of our time, due to the machine-made tools, rivets and a large variety of rolled shapes.

Transition. — While the work of the smith as a producer of art degenerated, the smith as a producer of tools for the shop and field improved, and now since the introduction of machinery that calls for hardened steel parts and for drop and pressed forgings, the smith with his knowledge of metals and how they should be heated has in the main passed to a work of far greater practical value, though not so artistic

CHAPTER II

IRON AND STEEL

THE purpose of this chapter is briefly to explain the production of iron and steel and to point out some of their characteristics. It is thought best not to introduce a theoretical discussion of metallurgy.

KINDS OF IRON

Iron is an element. As used commercially, it is never entirely free from impurities, some of which have a useful influence and others a harmful one. The presence of useful impurities in the iron and the method of obtaining the metal from the ores gives rise to three general classes of iron: i.e., cast iron (pig iron), steel and wrought iron.

There are two characteristics common to all of the above classes, i.e., all contain iron to the extent of 92% or more, and all contain the element carbon, as the next most important constituent.

Pig Iron is the raw form of iron just as it comes from the blast furnace. Almost all iron and steel are reduced from the ore to the form of pig iron, and are then refined by various processes into the cast iron, steel or wrought iron.

Cast Iron is the most impure form in which iron metal is used. It is weak and brittle; it cannot be heated and forged. To be shaped it must be melted and cast in molds, or machined with cutting tools. The per cent of impurities in cast iron varies between wide limits as can be seen from the following table:

Analyses of Cast Iron [1]

G. C.	C. C.	T. C.	Si.	Mn.	P.	S.
2.73	0.66	3.39	2.42	1.00	0.31	0.04
2.83	0.79	3.58	1.59	0.79	0.485	0.08
		3.75	0.85	0.50	0.45	0.05

Steel. — Steel is the name given to various compounds of iron and small quantities of carbon, silicon, manganese, sulphur, and phosphorus. It is purer and stronger than cast iron and can be shaped either by being melted and cast into molds or by being forged.

Special steels contain, in addition to the impurities mentioned, definite proportions of chromium, tungsten, manganese, nickel, vanadium, and wolfram. In general, steel is classified according to its amount of carbon, as follows: soft steel, having less than 0.3 %; medium steel, having from 0.3 % to 0.75 %; and hard steel, having from 0.75 % to 1.50 %.

Wrought Iron. — Wrought iron is the purest commercial form of iron. It is similar to very low carbon steel, except that it is not produced by being melted and cast in molds, and that it is always forged to the desired shape. It rarely contains more than 0.12 % of carbon. It is produced by the refining of pig iron by a process known as puddling.

Harmful Impurities. — The impurities in iron and steel that are harmful are sulphur, arsenic, and phosphorus. Sulphur causes the metal under the hammer to crack or

[1] G. C. is Graphite Carbon, C. C. Combined Carbon, T. C. Total Carbon. The other terms have their regular chemical significance.

crumble, when worked hot, while phosphorus causes it to crack or crumble when worked cold.

Hot Short or Red Short. — When impurities, such as sulphur and arsenic, render the metal unworkable at a red heat, it is said to be *red short*, and when unworkable at a welding heat it is *hot short*.

Cold Short. — The phosphorus impurity will cause iron to crack when it is worked or hammered while cold. Iron in this condition is said to be *cold short*.

FUEL — FLUXES AND ORES

Iron is mined in the form of oxides or carbonates. All of these ores generally are mixed with earthy and other impurities in widely varying proportions, and must be reduced by means of heat, thus making the use of various fuels and fluxes necessary.

Fuels. — The fuels which are used for reducing iron ores are, chiefly, as follows: charcoal, anthracite, and coke. The purer these fuels are the better will be the iron produced, other things being equal.

Fluxes. — A flux is a substance which is added to metalliferous bodies and unites with the foreign matter, to form a slag which is fusible. The flux to be employed will vary with the nature and amount of the impurities in the ore. Thus iron ores containing silicate of alumina need a flux of lime. Fluor spar is used as flux for sulphates of barium, calcium, and strontium.

Ores. — The ores which are used chiefly in the production of iron are the oxides of iron. They are here briefly described.

Magnetite (Fe_3O_4) when pure contains 72.41 % of iron. It is the richest and purest iron ore. Swedish iron is made from it.

Red Hematite (Fe_2O_3) is a rich, red-colored ore containing 70 % of iron. It is the most plentiful and widely distributed iron ore, and is the principal ore from which bessemer steel is made.

Brown Hematite ($2\ Fe_2O_3 + 3\ H_2O$) is red hematite chemically combined with water. When pure it contains 59.97 % of iron.

Calcining or Roasting. — Before iron ore can be used as iron, it must be reduced to the metallic state by the process of smelting. Ores, such as the *carbonate* and *sulphide* ores, which contain much volatile matter, are calcined, i.e., heated slowly to a temperature below that of fusion, in order to drive off the volatile matter. In this manner the carbonate ore, Fe_2CO_3, is changed to Fe_2O_3. The sulphide ore, FeS_2, is acted upon in a similar manner.

The calcining is performed by placing the ore, together with the proper amount of fuel, in heaps in the open air or in roasting kilns. The fuel is ignited and the whole mass gradually heated.

Water is driven off as steam, and the carbon dioxide (CO_2) and sulphur (S) go off as gases. Roasting in kilns is more satisfactory than in heaps, since the kilns can be worked continuously, for they are shaped like a large foundry cupola.

REDUCTION AND REFINING OF ORES

The Blast Furnace. — The first operation in the refining of iron ore is performed in the blast furnace (Fig. 10). The product is called either cast or pig iron. The furnace consists of a vertical shaft of iron or steel plate from 40 to 100 feet high, — the standard height ranging from 75 to 85 feet, — which is lined with a refractory material, thus leaving an interior circular space from 12 to 30 feet at its largest diameter, which maximum diameter is just below mid-height. From this region, the walls contract gradually in both upwards and downwards.

The lower sloping portion is called the boshes. This terminates at the top of a cylindrical portion called the crucible, the bottom of which is called the hearth.

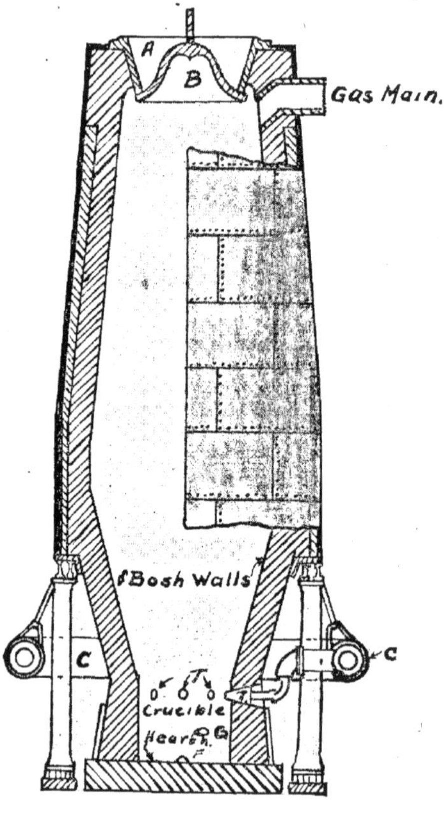

Fig. 10

The furnace is filled by placing the fuel, ore, and flux in the hopper (*A*), and then lowering the bell (*B*). Alternate layers of fuel, flux, and ore fill the furnace to the top of the melting zone.

Air is supplied under pressure of from fifteen to twenty-five pounds per square inch through a blast main or bustle pipe (*C*), which encircles the lower part of the furnace (Fig. 10). Smaller pipes connect the bustle pipe with the interior of the furnace through *tuyeres* (*T*), near the top of the crucible. The air passes up through the furnace, thus supplying oxygen for combustion. The gases pass off through a gas main at the top of the furnace, later to be used in the hot-blast stoves, for power in engines, or under boilers. The metallic iron and slags[1] both descend as liquids to the bottom and accumulate in the crucible (the light slag on top) until they are tapped off. The iron is tapped off at intervals at (*F*) and the slag at the cinder notch (*G*). The action of the blast furnace is as follows: As the fuel is consumed at the bottom the charge moves slowly downward and its temperature is constantly increased. The ore becomes roasted into either Fe_2O_3 or Fe_3O_4. The limestone is changed to lime by giving up carbon dioxide (CO_2), and the fuel is burned either to carbon monoxide (CO) or carbon dioxide. This CO_2 formed in these reactions, coming in contact with the fuel, is decomposed into CO and O. The lime unites with the silica and alumina, forming a double silicate of aluminum and lime, which is melted in the lower part of the furnace. The iron oxide (Fe_2O_3) freed from its impurities, coming in contact with the carbon monoxide (CO), gives up its (O), forming carbon dioxide (CO_2), and free iron. The iron, as it passes down into the hotter parts of the furnace, is in a spongy condition, in which state it readily absorbs carbon from the incandescent fuel. By this carburization, the melting point of the iron is lowered so that it becomes melted and runs to the crucible as cast iron, and there accumulates till drawn off. The slag flows toward the bottom of the furnace, but being lighter than the iron, floats on it and is tapped off at a higher level.

The air blast, before it reaches the furnace, is heated by passing through the hot-blast stoves (Fig. 11), which are heated by the gases from the furnaces. The stoves are used in pairs, so that one can be heated while the other is heating the blast.

Wrought iron, the purest commercial form of iron, is

[1] The substance resulting from the fusing together of the flux and the impurities in the ore.

made by decarburizing pig iron by a process called *Puddling* in a furnace called a Puddling Furnace (Fig. 12). Such a furnace consists of a horizontal concave hearth *B*, covered by a low arched roof, which reverberates heat upon the iron to be refined, which heat is produced by the combustion of a gaseous fuel in the space between the roof and the hearth. This results in burning out nearly all the carbon, silicon, and manganese, and some of the phosphorus and sulphur.

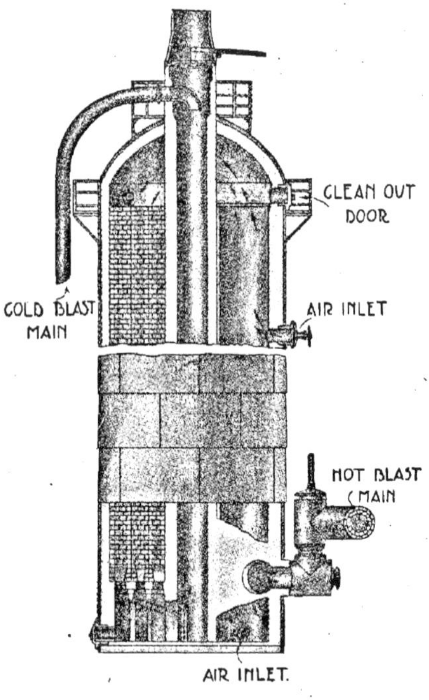

Fig. 11

There are two puddling processes, the dry and wet. In each, the operation is composed of three periods — fusion, robbling,[1] and forming the blooms.

Dry Puddling. — In this process, white or refined iron is chiefly used. The charge consists of about 4 cwt. of metal and some rich slags. This is partially melted, in about half an hour, to a pasty mass. The mass is then

[1] Robbling is the moving of the mass by means of the tools of the puddler.

stirred with an iron bar in order to expose all of the parts to the oxidizing influence of the air. As the impurities are removed, the iron becomes less fusible, and requires that the temperature be greatly increased to keep the iron liquid, but as it is not, the particles begin to solidify. The particles of iron are worked together into balls, weighing about 80 lbs., by the "puddlers," with a "puddle bar."

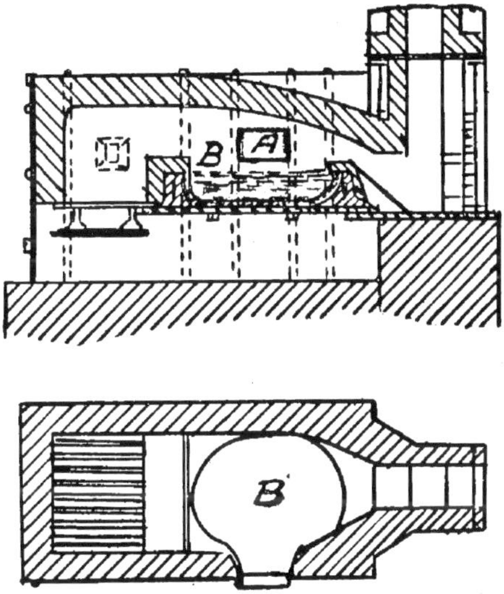

Fig. 12

Wet Puddling. — Wet puddling has succeeded the dry method, as the preliminary refining is thereby dispensed with. The pigs used in wet puddling are siliceous or strongly carburized.

The bed and sides of a modern puddling furnace are lined with refractory materials, such as mill scale or rich ore, which are rich in oxygen. When the iron is melted, it is acted upon by two oxidizing influences, the air and the iron oxide. The operation is shorter and the product more uniform than in the dry puddling process, since the robbling is more vigorous.

The slags thus formed take up oxygen from the air, causing FeO to change to Fe_3O_4, and this oxidizes the impurities in the iron, silicon and manganese being acted upon first, then phosphorus, sulphur, and carbon.

While carbon is being oxidized, carbonic oxide is formed, which bubbles to the surface, thus producing what is called the boiling stage. The entire mass is in a state of violent agitation. As the impurities are removed the iron gradually comes to nature, or solidifies, and is worked into balls as in the process of dry puddling. These balls are sponge-like masses of wrought iron, the interstices of which are filled with liquid slag. After they are taken from the furnace, they are raised to a welding heat, the slag is squeezed out, and the metal welded into blooms with a hammer or squeezer. Then they are passed through rolls and shaped into merchant bars.

MILD STEEL

Open Hearth Process. — In this process pig iron is melted and to it is added wrought iron or cold steel scrap. The furnace used is the *Sieman's Regenerative Furnace* (Fig. 13). It is a gas-fired furnace, using either natural or producer gas, which is made to give a very much higher temperature by being heated or regenerated before burning. The furnace has a hearth (*A*), covered over with a low flat roof to reverberate the heat. There are at least two passageways on either side, one at (*R*) for air and the other at (*S*) for gas, each connecting with a separate regenerator filled with fire-brick, which are laid with small openings between them to allow the passage of gas and air. The gas and the air enter the furnace flues on the same side of the hearth, each going through its own regenerator compartment to the terminals, where they mix and burn over the hearth in the combustion chamber (*B*). The hot gases of combustion pass out through the passages on the side opposite to that of the other set of regenerators and in so doing heat the checker work (as the bricks are called) to a very high temperature. When one side has become hot and the other has cooled, the flow of gas and air is reversed,

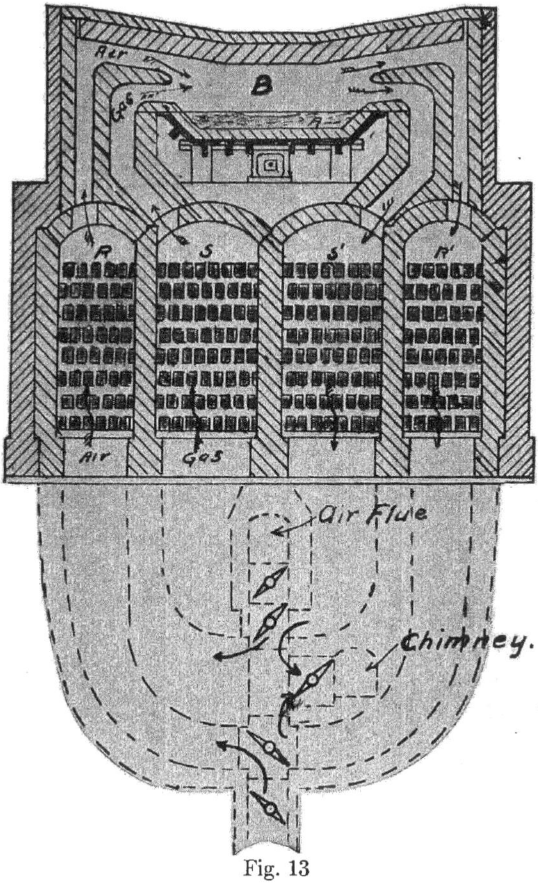

Fig. 13

so that the entering gas and air become highly heated in passing through these hot regenerators, and in burning produce a very high temperature. The spent gases pass out through the opposite set of regenerators and heat them ready for another reversal.

The Siemans-Martin, or pig or scrap process, consists of diluting, in a regenerative furnace, pig iron with scrap wrought iron or steel. The intense heat of the furnace keeps the steel liquid until it is poured into molds. There are three methods of producing open hearth steel.

By the *first method*, the pig iron is charged on the hearth and melted with an oxidizing flame, during which process part of the silicon, manganese, and iron combines with the oxygen of the flame, as in the puddling process. Scrap is then charged in, but slowly, in order not to chill the melted metal. The scrap, having but little carbon, reduces the percentage of carbon in the melted total, which is converted into mild steel or essentially wrought iron, kept liquid by the intense heat. The injurious FeO that is present is eliminated by adding ferro-manganese, the manganese uniting with the oxygen to form slag. The resulting melt is cast into molds.

The *second method* consists in charging wrought iron puddle balls into the melted pigs.

The *third method* consists of charging the pig and scrap together.

The *Siemans, or pig and ore process.* By this method, a bath of cast iron is decarburized by adding ore rich in oxygen, as Fe_3O_4 or Fe_2O_3. The oxygen unites with the carbon, silicon, and manganese of the cast iron to form slag and CO.

In the process just described, the lining of the hearth of the furnace is of silica which acts on the phosphorus in the iron, changing it to phosphoric acid, which remains in the iron. If the pig iron contains much phosphorus, a lining of lime and magnesia is used. When iron is melted in a furnace having a lining of lime and magnesia, the process and furnace are said to be basic; and when a silica lining is used, they are termed acidic.

Bessemer steel is made by a process differing radically from those described above in that liquid pig iron is converted into mild steel, no fuel being used, other than the oxidizable elements in the iron, which elements are burned out by the forcing of air through the liquid

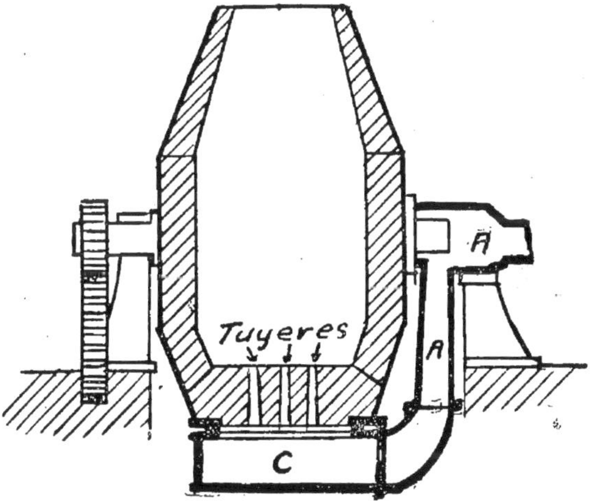

Fig. 14

metal. The operation is performed in a Bessemer Converter (Fig. 14). It consists of a steel shell, lined with a refractory material, and in shape resembling the hub of a wagon wheel. It has a double bottom which contains an air chamber (*C*), connecting with numerous small openings called tuyeres. Air enters through the blast main (*A*). The entire converter is hung on trunnions so that it can be turned to a horizontal position for charging and back to the natural for blowing.

To be *charged,* the converter is turned (Fig. 15) so that the liquid cast iron can be poured into it from a ladle or run into it direct from the blast furnace (Frontispiece). The converter is so made that when it is tipped for charging the iron cannot rise as high as the tuyere holes. Before the charged converter is turned to the vertical position, the air blast of about twenty-five pounds per square inch is turned on to prevent the iron from running out of the tuyeres while the converter is returning to, and assuming the vertical position.

When the converter lining is acid, the silicon, manganese, and carbon of the iron are burned out simultaneously by means of the air blast, which elements act as fuel to raise the temperature of the charge. The resulting slag is blown out of the mouth and is there burned, producing brilliant sparks. In the converter the carbon is burned to CO, which at the mouth burns to CO_2. Thus the cast iron is changed to wrought iron or mild steel, which is kept liquid by the intense heat. When the charge has been properly converted, and this is told by the color of the flame, the converter is again turned on its side and molten *spiegeleisen* is poured into the metal, to recarbonize it, to reduce iron oxide, and to remove gases. The converter is then turned farther down so as to dump the entire charge into a ladle. From the ladle it is poured into ingot molds. In the acid process, or when the converter has a silica lining, no phosphorus is removed from the iron, and so with this lining in use the iron must be low in phosphorus.

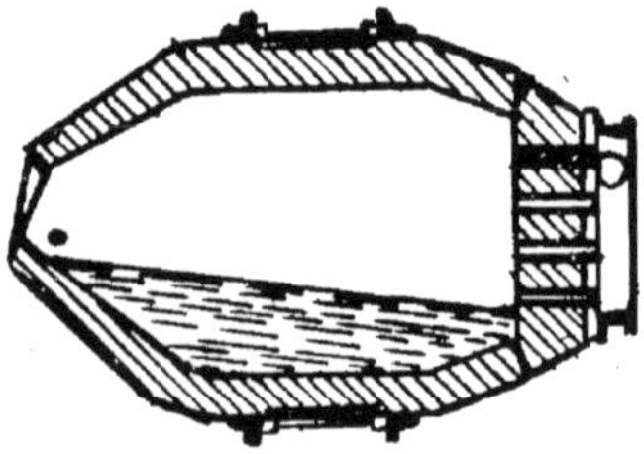

Fig. 15

Basic Process. — For iron with much phosphorus a basic lining is used. The phosphorus in this case takes the place of the silicon as the fuel. In the basic process we have so large a quantity of slag formed that it must be poured off, and then the converter is returned for a short period of blowing called an *after blow*, to convert the remaining phosphorus into slag. The remainder of the process is the same as the acid process. The temperature at which the metal is poured into the ingot molds must be carefully regulated or the ingots will be porous. If the cast iron has too much of the fuel elements the temperature will be too high; and if too little, it will be too low. When too high, the temperature is reduced by adding scrap steel; when too low, the converter is tipped to one side and a part of the tuyere holes are uncovered to introduce oxygen to burn the CO to CO_2 *inside the converter*, instead of at the mouth, and thus to heat the metal as in a reverberatory furnace.

The capacity of a Bessemer Converter varies from 1000 pounds to about 15 tons and the average time of working 10 tons is about 20 minutes.

Ingot molds are made of cast iron. They are from 6 to 7 feet high, open at the top and bottom and tapering from the base, the large end being the bottom. The bottom is closed by setting the mold on a cast-iron plate. After the molds have been poured and allowed to cool slightly, the mold is lifted off and the ingots are placed in a soaking pit, where they can cool slowly so that the interior will solidify and the outside will not become cold. This gives a uniform temperature throughout, so that the ingot is ready for rolling into bars, rails, etc.

It is interesting to note that to make a steel rail by the Bessemer process, after the iron leaves the blast furnace no additional fuel is necessary other than that contained within itself.

Tool Steel or Crucible Steel. — The best tool steel is made from wrought iron by the *cementation process.* Bars of the purest wrought iron are cut into pieces about

$\frac{5}{8}''$ by 5″ by 12″ and are packed with alternate layers of finely crushed charcoal in boxes of fire-brick. The cover is luted on with clay to prevent the air and furnace gases from injuring the contents. It is then placed into a furnace and slowly heated to about 3000 degrees F. At this temperature it is held for several days and then allowed to cool. While thus heated the iron absorbs a portion of the charcoal (carbon), but not uniformly, the carbon being more dense near the surface. From the blisters that have formed on its surface, the product of this process is termed blister steel. These bars are coarse-grained and brittle and must be broken into small pieces, placed in fire-clay or graphite crucibles, and then placed in a crucible furnace where a temperature high enough to melt the steel is obtained. When melted, the crucibles are lifted out and the contents poured or *teemed*, as it is called, into ingots (about 3″ by 3″ by 3″) of the same composition throughout but coarse-grained and weak. It is next reheated and rolled or hammered into commercial bars of fine grain. The more the bar is hammered the closer and finer becomes the grain.

Cheaper grades of steel are made by omitting the cementation process and charging the wrought iron and charcoal into the crucible. When the iron melts it absorbs the carbon. A cheap grade also may be made by melting together wrought iron and cast iron.

The special high-speed steels are made in about the same way as the carbon steels, except that they contain amounts of other chemical elements or compounds added to the contents of the crucible.

Rolling Mill. — Wrought iron and steel, after they have reached the bloom or ingot stage, must be rolled. This rolling is done in a rolling mill which consists of two or three rolls held in position by a frame called a *housing*. A mill of the old form contained two rolls and was called a *two high mill.* In this mill the rolls

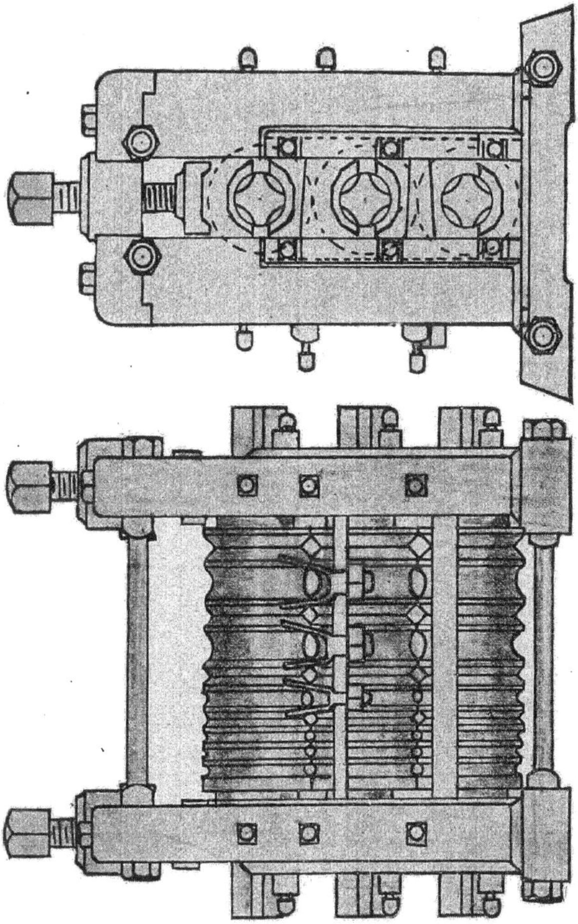

Fig. 16

revolved in opposite directions so as to draw the iron into them. The rolls are either plain, as when the plates are being rolled, or grooved, as when *shapes* are to be made.

In the two high mill after the bars had been passed through the roll, they had to be passed back to the starting side by being lifted over the top roll. They were then passed back through the rolls, but through a smaller and possibly differently shaped groove, until of desired shape and size. This passing of the iron back over the top roll consumed time and allowed the iron to cool; hence there was made the *three high mill* (Fig. 16) in which the iron could be reduced by rolling through the two top rolls on its way back. Now a train or several series of rolls are used so that a bar can be started in at one end of the series of rolls and come out finished at the other. Each series of rolls must be speeded enough faster than those before so that they will pass the lengthened bars through without allowing the bars to buckle.

The hammering of tool steel is accomplished by means of ordinary steam or power hammers.

REVIEW QUESTIONS ON IRON AND STEEL

1. What is iron obtained from?

2. Name the important iron ores.

3. What is calcining or roasting? How is it done? What does a roasting kiln look like?

4. What is a blast furnace used for? Describe its parts. How is ore and flux charged on? What is flux? What is the operation of the blast furnace? How and when is the charge tapped off?

5. What is a hot-blast stove? What is it used for? Why are hot-blast stoves worked in pairs?

6. What is wrought iron? What is puddling? Describe a puddling furnace. How many puddling processes are there? Name them. What kind of flame is used in dry puddling? What becomes of the impurities in the iron? What happens to the melting point of iron when the impurities burn out? What is a puddle ball; a puddler? Are the puddle balls solid? What is a bloom?

7. In wet puddling how is the oxygen obtained? What kind of lining is used? What advantage is there in using the wet process?

8. What is melted and made in the open-hearth process? What kind of a furnace is used? What are regenerators? What is the Siemans-Martin process? Describe the various methods. What does ferro-manganese do when added to the melt on an open-hearth furnace? What is the Siemans pig and ore process? Where does the oxygen come from in this process? What is the effect of the silicon lining of a furnace on phosphorus in the iron? Why is the hearth of the furnace sometimes lined, previous to melting, with alkaline substance as lime and magnesia? What is meant by basic process? What is the capacity of an open-hearth furnace?

9. What is Bessemer steel? How is it produced? How does it differ radically from other processes? How is the metal prevented from running out through the tuyeres? What furnishes the fuel? What does the spiegeleisen do? When is the lining of a converter acid or basic? Why must the temperature be carefully regulated during the pouring of ingot molds? Describe ingot molds. Why are ingots placed in soaking pits?

10. What is the first step in making the best tool steel? What materials are used? Why does the blister steel have to be remelted? What improves the grain of the steel, after it is poured into ingots?

11. What is a two high mill? A three high mill? What is a train of mills? Why must each set of rolls in a train have a successive different speed?

CHAPTER III

EQUIPMENT

After we have studied the nature and manufacture of iron and steel and before we take up the shaping of these materials into articles of ornament and daily use, it seems natural that the shop equipment used for this

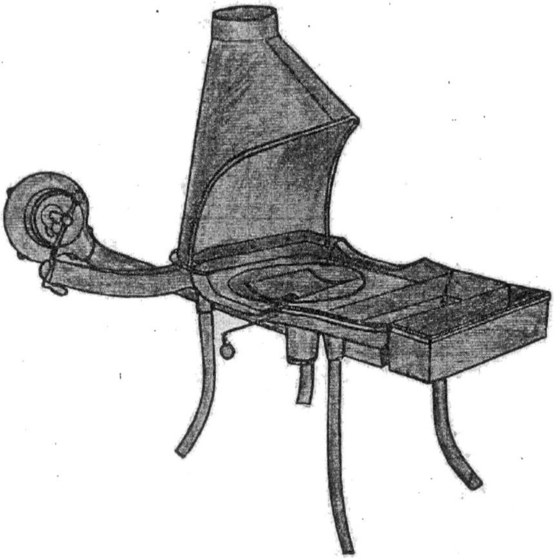

Fig. 17

purpose should be considered. This equipment may be divided into two classes: (A) general tools; (B) small or hand tools.

(A) GENERAL TOOLS

Forges. — The forge in which the iron is heated is the most important part of the equipment of a shop.

Forges range in size from a small portable iron forge to the large, stationary brick forge. Forges are of two types: the up-draft, and the down-draft.

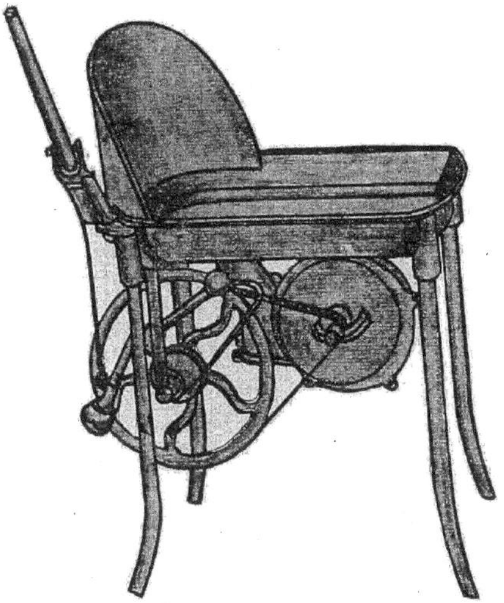

Fig. 18

In the up-draft forge, the smoke and gases pass up a chimney by natural draft (Figs. 17 and 18). Fig. 17 shows a portable forge having a blower operated by a crank. Fig. 18 shows a forge that is similar, except that the blower is worked by a handle which operates a ratchet and wheel. Fig. 19 shows the old-style brick forge which is blown by a bellows.

Fig. 19

In the down-draft forge, the smoke and gas are sucked or drawn under the hood and downward by a suction blower. Fig. 20 shows one of the several types of this forge. The downward draft forge is supplied with blast from a power-driven fan or blower, while a second blower is used to suck the smoke and gases under the hood and down and out of the room. Down-draft forges are the most desirable for schools or shops where several forges are used at one time, as they act to ventilate the shop by carrying off the smoke

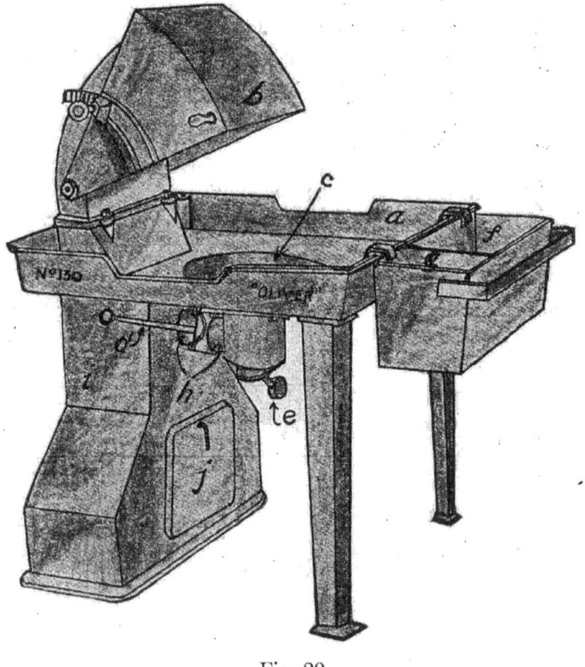

Fig. 20

and gases and to keep it cool by causing a circulation of air.

The parts of a forge are (*a*) fire-pan, (*b*) hood, (*c*) tuyere, (*d*) lever to regulate the blast, (*e*) lever to dump ashes and clinkers, (*f*) coal box, (*g*) water box, (*h*) blast pipe, (*i*) gas and smoke outlet, (*j*) cleaning door.

Fig. 21 shows two styles of tuyeres. That shown at (*a*) is a square plate with several holes and is connected with a lever so that to expel the ashes it can be dropped.

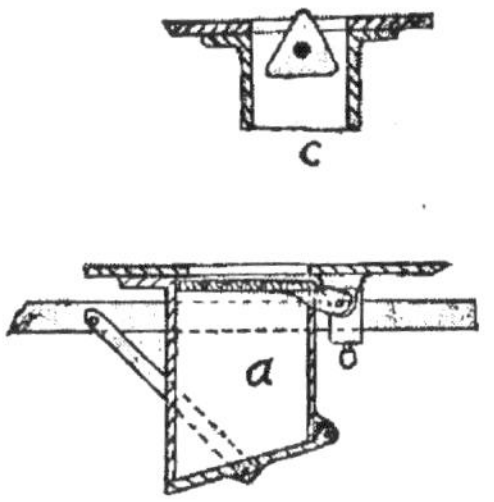

Fig. 21

Fig. 22

The lower door is a solid plate and is used to close the ash chute. The tuyere at (*b*) is so arranged that it can be revolved to make a wide or narrow fire. When in the position shown, the blast is spread and a large, wide fire results. When placed with the vertex of the triangle up, as at (*c*), the blast is converged to produce a narrow fire. At (*d*) is shown the "whirlwind" tuyere. The makers claim for this tuyere that it is anti-clinkering; that it

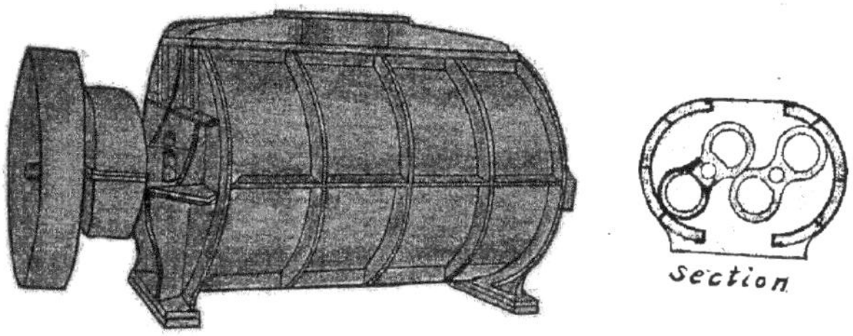

Fig. 23

produces a circular, rotary whirlwind blast; and that by the deep nest the heat is concentrated instead of being blown out of the chimney, thus making quickly and cheaply a hot and non-oxidizing fire. The hoods on down-draft forges can be raised and lowered.[1]

Blowers are of two types, the fan (Fig. 22) and the pressure (Fig. 23). The pressure blower is the better since it gives a steady blast, whereas that from the fan is spasmodic.

Fig. 24

Bellows (Fig. 24) were formerly much used to furnish the blast for blowing the fire, but they are rarely found in modern shops.

Anvil. — Next to the forge the anvil (Fig. 25) is the most important part of the shop equipment. Anvils are rated by their weight; No. 150 meaning one that weighs 150 pounds. The size used in shops ranges from No. 150 to No. 250 and is selected according to the work

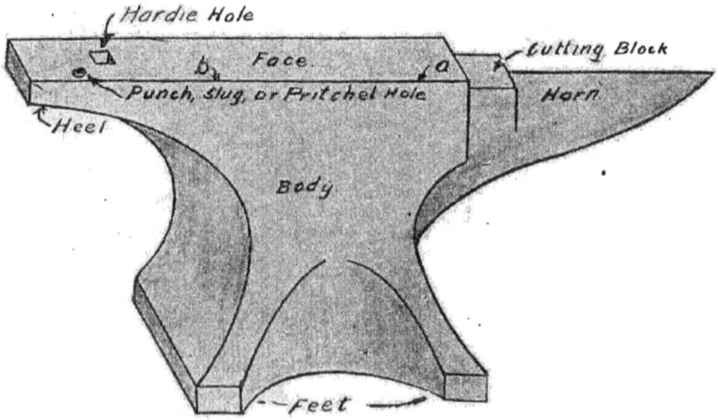

Fig. 25

to be done. Anvils weighing 100 lbs. are very satisfactory for school use. On the older anvils of English make the weight is stamped on the side. Three numbers are used;

[1] The author believes a forge open underneath is the better, as the parts are more accessible for repairs.

the first represents hundredweight (cwt.) of 112 lbs., the second or middle number, the quarters of a hundredweight, and the third or right-hand number, the odd pounds. Thus 1–2–8 means 112 lbs. + 56 lbs. + 8 lbs. = 176 lbs. This anvil would be called a 175 lb. anvil.

The names of the parts of an anvil are shown in Fig. 25. About half the length of the edge (*a*) is rounded to a quarter-circle as shown in Fig. 26. The radius (*r*) of the curve varies from about $\frac{1}{2}''$ at the cutting block to zero or a sharp edge at (*b*), Fig. 25. The balance of this edge and all of the other edges of the face are sharp and will cut stock when it is hammered against them. The object of this rounded edge is to have a place where stock can be bent at an angle without danger of being cut, as this might cause the work to crack. The square or hardie hole is designed to hold the hardie and bottom swages and fullers. The round hole, called the punch or pritchel hole, is used in punching holes, to give a place through which slugs can pass. The pritchel hole is also sometimes used as a heading tool. The face of the anvil is made of tool steel welded to the body, and hardened, so it is not easily injured by hammering upon it. The cutting block is left soft so that cutters and chisels when coming in contact with it when cutting or splitting through a piece of iron will not be dulled. Pieces to be cut should be placed on the cutting block and not on the face of the anvil.

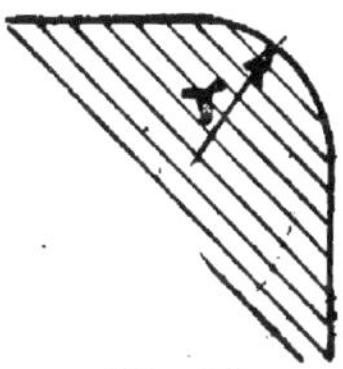

Fig. 26

The anvil should be set on a heavy block of wood or on a cast-iron base made for the purpose. The wood block is to be preferred since it is more elastic. The cast-iron base is neater in appearance and for light work is very good. The height of an anvil should be suited to the smith. For most convenient use it should be high enough, so that when he stands erect with arms hanging naturally the knuckles will just touch the top edge of the face. It should be firmly fastened to the block by straps across the feet or by other suitable devices (Fig. 27). There are other styles of anvils than that shown in

Fig. 27

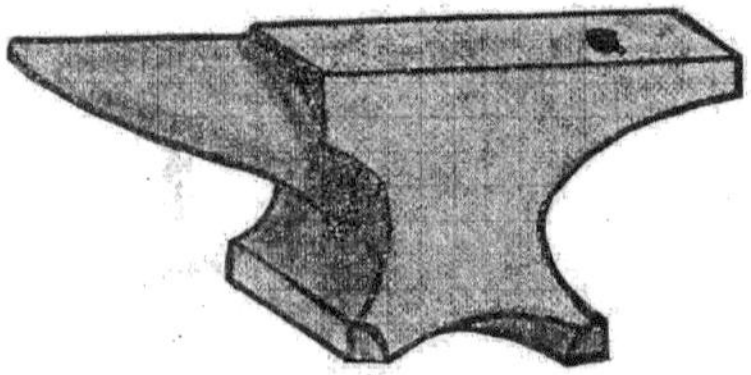

Fig. 28

Fig. 25; as the ferriers (Fig. 28), and the double horn (Fig. 29). They are used little and need no further reference.

Power shears (Fig. 30) and power hammers (Figs. 218–222) should be in every shop. Almost any size of shears can be had, but one that cuts stock from 1″ to $1\frac{1}{2}''$ square is large enough for most shops. Power hammers can be obtained that are rated from 25 lbs. up. The average work to be handled will indicate the size that should be selected.

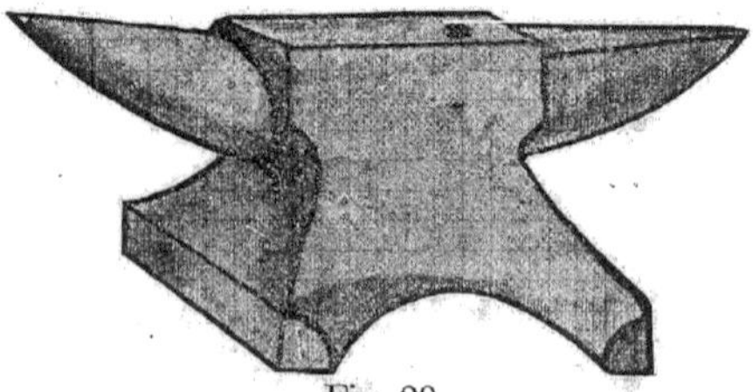

Fig. 29

Swage-blocks (Fig. 31) are used for many purposes but mainly in place of bottom swages. This tool consists of a cast-iron block pierced with numerous holes — round, square, and rectangular — provided with edges with grooves of various sizes, in circular and V-forms.

Fig. 30

The block is used either as a bolster or heading tool when lying upon its side, or as bottom swages when standing on one edge. The swage-block serves its purpose best when mounted on a stand (Fig. 32). The top (*a*) is made of 2″ by 2″ angles and bent to the size and shape of the block and mounted on four stiff legs. The lower leg of the angle (*a*) is on the inside, where it forms a shelf on which the block can be rested, as shown by the full lines at (*b*).

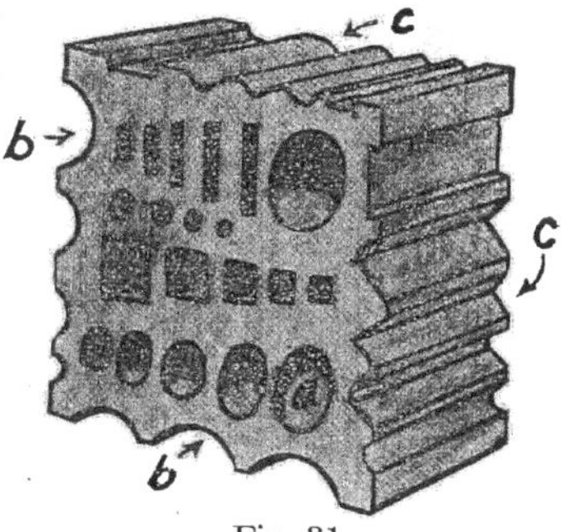

Fig. 31

The stirrup (*c*) is attached to (*a*) at the middle so that also the stand will hold the block in the position indicated by the dotted lines (*d*).

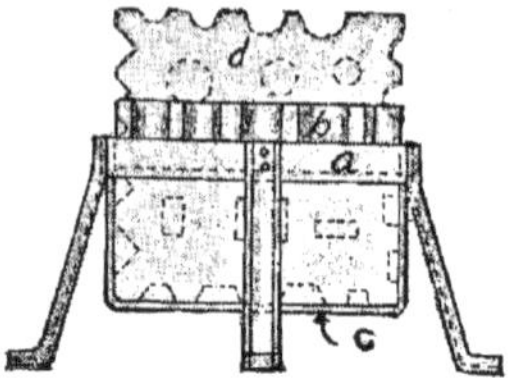

Fig. 32

The **Mandrel** (Fig. 33) is used for finishing rings to circular shape.

Bench. — A shop is not complete without a suitable bench on which to place a vise and to do laying out and other work.

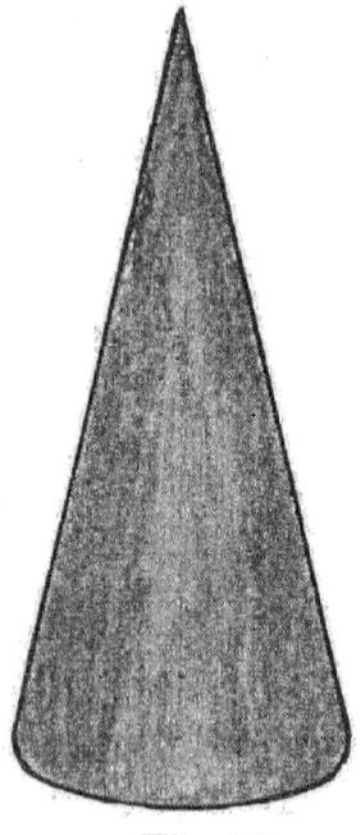
Fig. 33

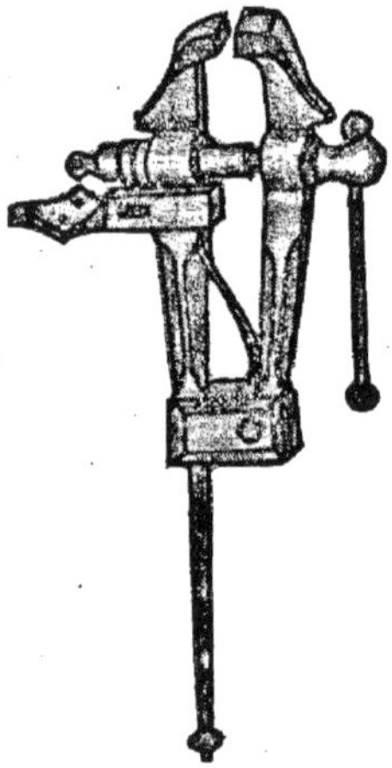
Fig. 34

Vise. — Fig. 34 shows a vise, usually called a blacksmith's vise, which is cheap and suitable for rough work, but the one shown in Fig. 35, having swivel jaws and an attachment for holding pipe, is to be preferred.

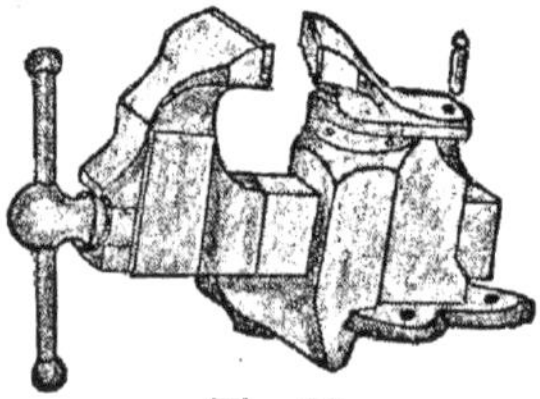
Fig. 35

Drill Press. — A smith's shop ought to have a drill press of some form. If the work is light, it may be a small hand-driven post drill, but a power-driven drill is to be preferred.

(B) SMALL OR HAND TOOLS

Hammer. — Fig. 36 shows four types of hammers used in the forge shop. The one most common is the ball pene shown at (*a*). The square-faced hammer shown at (*d*), known as the blacksmith's hammer, is very much used. The straight pene (*b*) and the cross pene (*c*) find favor on some classes of work and with some smiths.

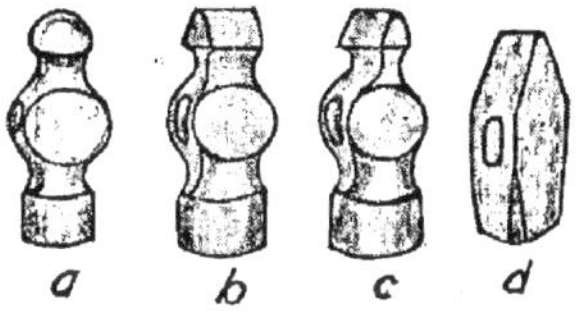

Fig. 36

The heads should weigh from $1\frac{1}{2}$ lbs. to 2 lbs. The length of the handles varies with the weight of the heads, but it should be about 14″ to 16″ long. The handle should be fitted to the head with great care so that the hammer will hang true; that is, so that the center line of the head will pass through the major axis of a cross-section of the handle and also be at right angles to the length of the handle (Fig. 37). If the handle is not fitted correctly it is impossible to strike accurately.

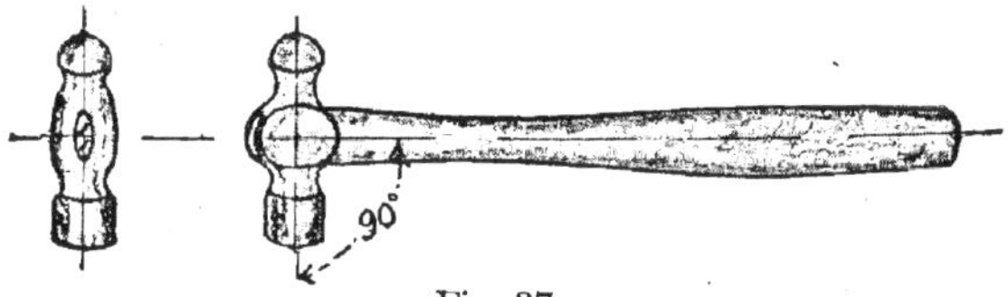

Fig. 37

A **Sledge** is a hammer weighing from 5 lbs. to 20 lbs. with a handle from 30″ to 36″ long. A sledge weighing from 8 lbs. to 10 lbs. is the best for ordinary use. It is used by the helper when heavy blows are needed either on the work direct or on some tool as a swage, fuller, or flatter. Fig. 38 shows a cross pene (*a*), straight pene (*b*), and double-faced sledge (*c*). The cross pene is used most; while the double-faced is found in but few shops. The eye for the handle in the straight and cross pene sledges is placed above the center of gravity, so that the sledge will naturally hang face down in the striker's hands.

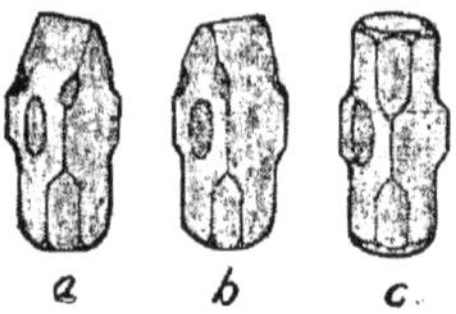

Fig. 38

Tongs take first place among the small or hand tools used by the blacksmith, for without them he could not hold the hot iron. Fig. 39 represents the flat or plain tongs used for holding flat iron. They should be made

Fig. 39

Fig. 40

with a groove down the center of the bit so that round stock also can be held in them. Fig. 40 represents box tongs used for holding rectangular stock which is to be bent or worked on the edge. The box shape

Fig. 41

Fig. 42

prevents the work from slipping around. Fig. 41 shows a pair of box tongs that can be adapted to a wide range of work by making several box pieces (*a*) to fit the various sizes of stock handled. Fig. 42 shows these tongs in use. Fig. 43 indicates a pair of round bit tongs for round stock. Fig. 44 represents the hollow

Fig. 43

Fig. 44

bit or bolt tongs suitable for holding stock having an end larger than the body of the work, as does a bolt. The bit can be made with either round or square grooves. The square ones are to be preferred for they hold either square or round stock. Fig. 45 shows tongs

Fig. 45

Fig. 46

for large round, or square work. Pick-up tongs (Fig. 46) are used, as the name indicates, for picking up work and tools from the floor. Fig. 47 shows pincer

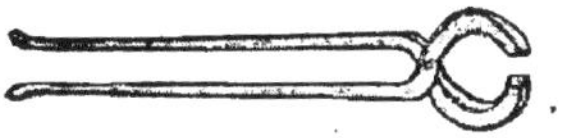

Fig. 47

Fig. 49

tongs. Fig. 48 represents tongs of special shapes with bent and crooked bits. In Fig. 49 is shown the proper way to grip flat work with plain tongs. The stock touches the bit at both (*a*) and (*b*). Fig. 50 shows the shape of bits for properly gripping round work and Fig. 51, the proper grip by means of the square bit.

Fig. 50

Fig. 51

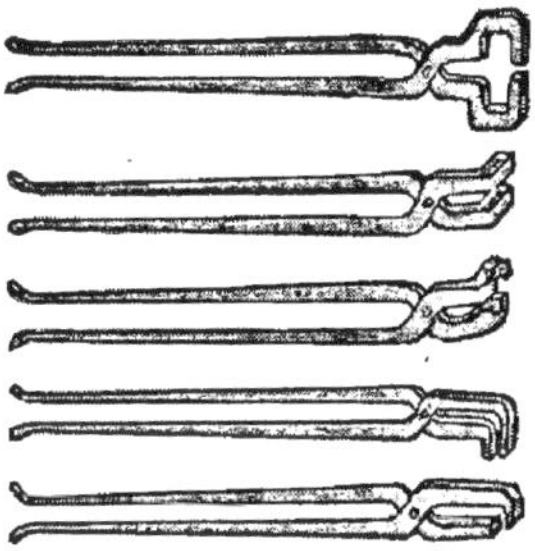

Fig. 48

Chisels.—Next in importance are the chisels used for cutting off or splitting work. Fig. 52 represents a hot chisel used for cutting hot stock. It is made thin and sharp so that it will penetrate the hot metal rapidly and thus prevent the loss of temper. It is thin, since great strength is not required to cut hot metal. Fig. 53 is a cold chisel or cutter, for cutting cold stock. It is made blunt for

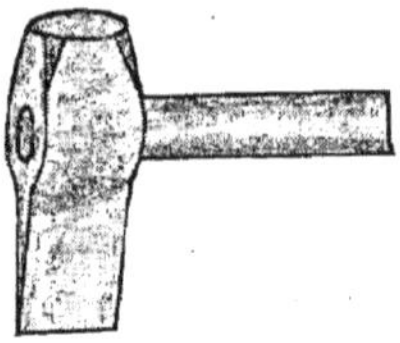
Fig. 52

Fig. 53

Fig. 54

strength. Fig. 54 shows a gouge chisel used in making round corners. It is an inside or an outside tool according to the way it is ground.

Punches (Fig. 55) are used for making holes of round, square or elliptical cross-section through stock. Fig. 56 indicates a bob or counter punch used for counter-sinking holes for screw heads. In Fig. 57 is shown a cupping tool used in rounding off or finishing the heads of rivets.

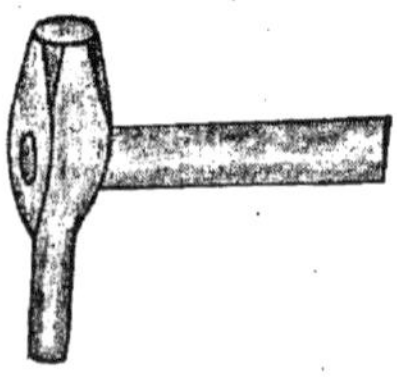
Fig. 55

Fig. 56

Fig. 57

The **Set Hammer** (Fig. 58) is used for setting down work and working in small places, or for producing sharp edges. It is usually made with square edges, though some have rounded ones and are then called round-edge set hammers. The *flatter* (Fig. 59) is used for smoothing work and giving a finished appearance by taking out the unevenness left by the hand hammer. Fig. 60 shows a special tool called a foot tool, used on work that cannot be reached by the set hammer. The foot reaches the part of the work that it is desired

Fig. 58

Fig. 59

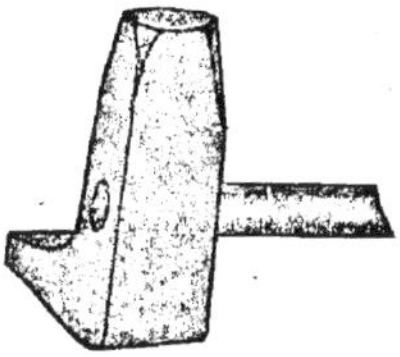

Fig. 60

to have worked, while the head remains where it can receive the blows from the sledge.

Swage. — Fig. 61 represents a top swage for rounding up work. Fig. 62 indicates the bottom swage. The bottom swage has a projection that fits into the hardie hole of the anvil. In Fig. 63 is indicated a spring

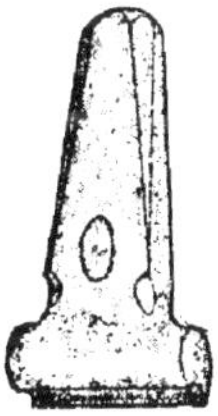

Fig. 61

Fig. 62

Fig. 63

swage. This consists of a top and bottom swage held together by a spring which insures the proper position of the two swages. Fig. 64 represents a nut swage for making nuts and bolt heads. A collar swage, used for truing up collars on shafts, is indicated in Fig. 65.

Fig. 64

Fig. 65

Fuller. — Fig. 66 shows a top and Fig. 67 a bottom

fuller, used for bending the fibers of iron, without cutting, when a section is to be reduced at some point.

Fig. 66

Fig. 67

Fig. 68

Fig. 69

Hardie. — A hardie, which is nothing more than a bottom cutting tool, is represented in Fig. 68. It fits in the hardie hole of the anvil and is used for cutting off stock, either alone or with the hot or cold cutters.

Fig. 70

Fig. 71

Fig. 72

Miscellaneous. — Fig. 69 shows an anvil cone; Fig. 70, a fork for bending work; Fig. 71, for bending flat stock; Fig. 72, a hand heading tool, for making heads on bolts; Fig. 73, a saddle for drawing out forked pieces; and Fig. 74, a mandrel for finishing nuts, etc.

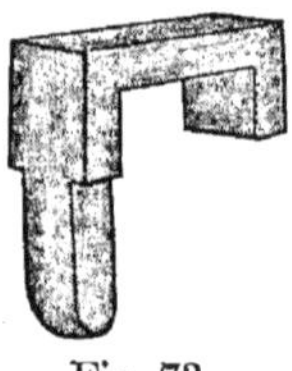

Fig. 73

Fig. 74

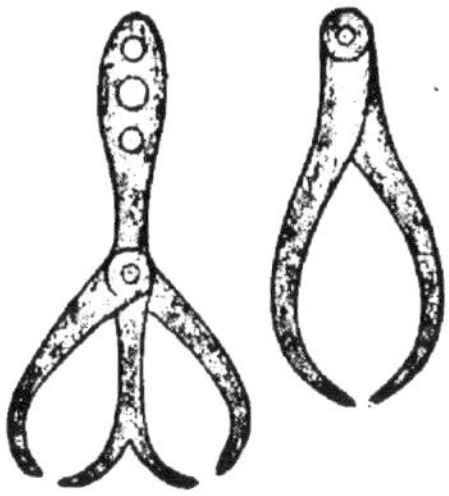

Fig. 75

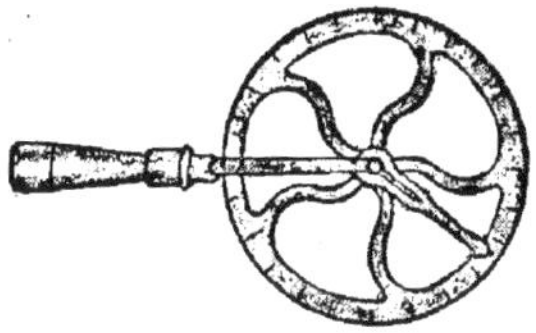

Fig. 76

The shop should be further provided with calipers (Fig. 75), a steel square, and a measuring wheel (Fig. 76).

QUESTIONS FOR REVIEW

Into what classes may a forge equipment be divided? What is the forge used for? What kinds are there? How many types of forges? How does each work? Name the parts of a forge. What type of forge is the most desirable for a large shop? Name the types of blowers, and tell which is the best. Name the parts of an anvil. How are they selected for size? What is the round edge for? How should a hammer head be fitted? Name the different types of hammers. What is the sledge used for? Name the types of sledges. Describe six types of tongs. What is a hot cutter? Why is its edge made thin? Why is a cold cutter blunt and stubby? What are the shapes of punches?

CHAPTER IV

FUEL AND FIRES

The selection of a fuel and the making and care of a fire are essential to first-class work. Especially is this so in welding iron and working tool steel. No matter how good his shop equipment or how fine his set of tools, the smith cannot produce good work with them if his fire is dirty, too thin, or otherwise not suited to the work in hand.

Fuel. — The coal employed in the forge should be a bituminous coking coal, as free from ash, sulphur, and other injurious matter as possible. It should be fine screening or run of mine. If it is run of mine, the large lumps should be broken into fine pieces. The excellence of coal is determined by the watching of how the fire burns. If it sometimes gives a hot fire and sometimes not; if it comes up fast and then rapidly dies out; if the flame is red, edged with blue; if the coke that is formed is dark-colored and easily crumbled; if it is difficult to make welds; then the coal is of inferior quality. The following tests, offered by the Pennsylvania Coal and Coke Company, can be performed by any one and will aid greatly in the selection of suitable forging coal.

Tests. — Take several pieces the size of your fist and crack them open. If little white scales or brown deposits appear between the layers, they are sulphur. It is bad for any iron and steel and absolutely prevents making good welds. A suitable smithing coal contains no such white scales or brown deposits.

Look at the coke formed around the edge of the fire. If it is not solid, and not of a clear gray color, the coal contains a large quantity of dirt. A suitable smithing coal forms a clear gray coke, of even grain, which, when burned, makes a hot, steady fire.

A blue edge around the flame indicates a large amount of injurious sulphur. A suitable smithing coal, being practically free from sulphur, makes a pure red and yellow flame.

Look closely at your coal pile and see how many pieces of dull gray slate you can pick out, just from the surface of the pile. Slate is not coal. It itself will not burn, and it keeps the coal with which it is mixed from burning freely.

If your fire is hot in spots, or for a short time, and then "drops out," the coal is low in heat efficiency and is not adapted to smithing. A suitable smithing coal maintains a high, clear heat for a remarkably long time, because it is all pure heat-giving material.

Charcoal is often substituted for coal, especially when tool steel is being worked. It is almost pure carbon and contains no sulphur, and will therefore introduce nothing injurious into the iron or steel. Due to its light weight a strong blast would blow it out of the fire. It therefore is poorly suited for work requiring a strong fire. Its use is rapidly growing less.

Fire. — Three types of fires are used by the smith; namely, the plain open fire, the side-banked fire, and the hollow fire. Each may be either oxidizing or reducing, according to the depth of the fire and the amount of blast. When the fire is so thin that the blast can pass through it without all the oxygen being consumed, this oxygen will attack the iron, forming ferrous oxide or scale. A fire of this kind is an oxidizing fire, and should not be used. On the other hand, a deep fire in which all the oxygen is consumed before reaching the iron is known as a reducing fire, for it sometimes extracts oxygen from the scale, changing the scale back to iron. It is therefore essential in whichever way the fire is built to keep the fire so thick that the oxygen will be consumed before reaching the iron.

The Plain Open Fire is the simplest and easiest fire to make. The unburned coal and coke from the last fire is scraped back from around the tuyere and the cinders and ashes are removed; thus a hole over the tuyere is left. A handful of shavings is placed in this hole and lighted. Some of the small pieces of coke from the last fire are now scraped over the shavings and the blower is started. In a very short time, white smoke arises, followed by tongues of flame. If the flame does not soon appear, a small opening is made through the center of the fire or coke by passing the poker down to the tuyere. This allows the entrance of air and the gas burns with a flame. As soon as the coke ignites, wet green coal is placed at the edge of the fire and patted down with the fire shovel. As the fire is used, green coal is added from time to time at the edge of the fire and the coked portion is pushed forward over the tuyeres. This kind of fire is useful for small work but has the disadvantage of spreading badly. Wetting the coal around the edges will keep the spreading down to some extent.

The Side-banked Fire is the best for general use. It is made in the following manner. The coke from the last fire is cleared away from around the tuyeres for a considerable distance, and the ashes and clinkers are removed. A block of wood wide enough to cover the tuyere, and as long and as high as the banks are to be made, is placed over the tuyere. The block should extend lengthwise away from the operator. Wet green coal[1] is now packed on each side of this block (Fig. 77) to a depth of four inches or more, according to the

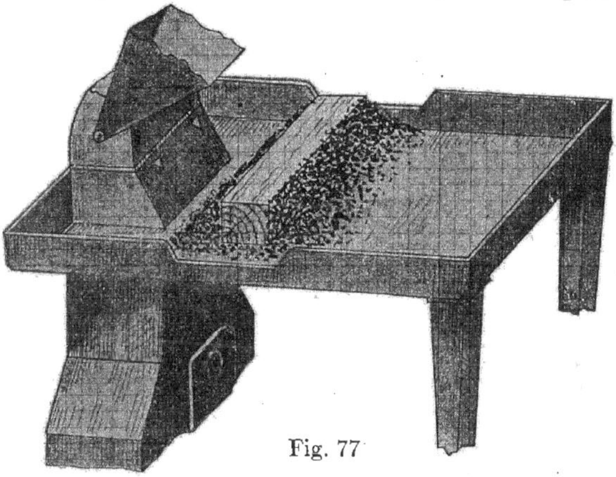

Fig. 77

length of time the fire is to be used. The block is removed and the banks nicely rounded with the hands. This leaves a trough-like space (Fig. 78) between the

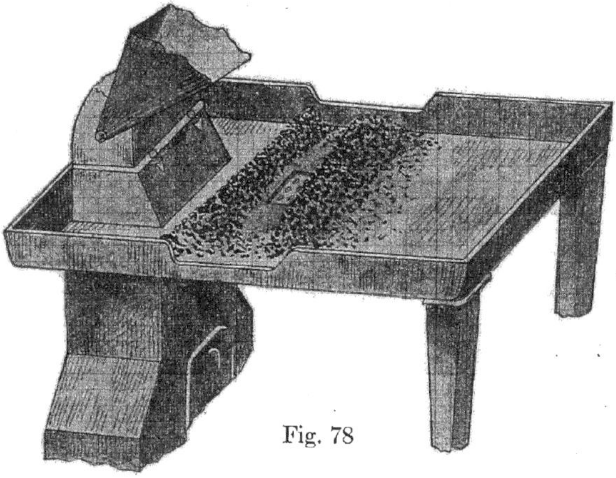

Fig. 78

[1] The fine green coal is wet with water and turned with a shovel until all is equally wet. Care must be taken not to get it so wet that the water will run out of it.

banks, which should be left open at the ends. The fire is started by placing shavings between the banks and igniting them. The fuel in this type of fire is the coke formed from the banks of the previous fire and broken to proper size and added to the fire as needed.

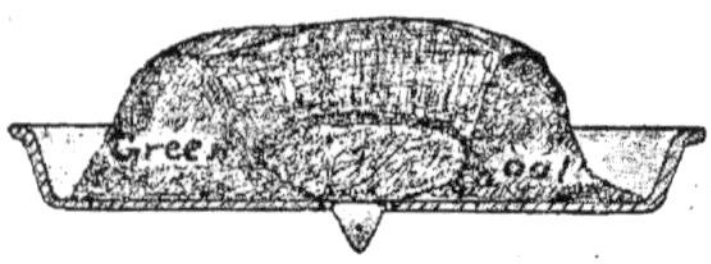

Fig. 79

The Hollow Fire is made in a manner similar to the side-banked fire. The sides of the fire are farther apart and are curved in and brought together at the back, making one continuous "C"-shaped wall (Fig. 79). The fire is now started inside of this wall and built up with coke, as high as the desired inside height of the opening. The roof is next laid in that the smith completely covers the coke with well-tamped wet green coal, from bank to bank, except for the opening in front, as shown by the transverse section of Fig. 80. As the coke burns down the inside of the roof is coked enough to bind the coal. This makes it self-supporting, and also leaves an open space between the roof and the fire, shown in longitudinal section of Fig. 81, in which articles can be heated.

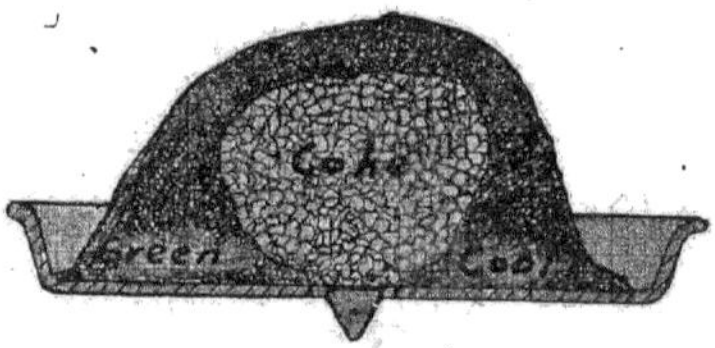

Fig. 80

Fig. 81

This fire is not very extensively used, but when it is, a very intense heat can be obtained as the heat is reflected or reverberated from the roof upon the metal which is being heated. An important use of this fire is in welding steel to iron.

QUESTIONS FOR REVIEW

Why is the selection of a fuel important? What is the most common fuel for forging? What are the tests for good coal? What do white and brown spots in coal indicate? What gives the blue flame sometimes seen in a forge fire? Why? What does the color of the coke indicate? What does slate in the coal pile do? What does a short-lived fire indicate? Is charcoal a good fuel? Why? How many types of fires are there? What is an oxidizing fire? What is a reducing fire? Describe how each type of fire is built. Which is the best for general use?

CHAPTER V

DRAWING DOWN AND UPSETTING

In this chapter, drawing down and upsetting, the two fundamental operations of forging, are treated. There is scarcely a forging of any description that does not involve one or both of these operations. A discussion of these processes, however, must be preceded by instructions: How to stand at the anvil, and how to hold the hammer and sledge.

Fig. 82

Position at the Anvil. — The smith should stand erect (Fig. 82) about 12″ behind the anvil. His feet should be at right angles, the heels not over six inches apart, with the left foot nearly opposite the center of the anvil. This will place the body almost opposite the heel of the anvil and the face turned a little towards the horn. The tongs should be carefully selected so that they will grip the work tightly. They should be held firmly in the left hand,[1] straight out from the center of the anvil, and at such a height that the work

[1] A right-handed smith is being considered in all these descriptions.

will lie flat on the face of the anvil. If the hand is held too high or too low, the work will become bent by the hammer blows. The hammer is held in the right hand, so that the handle will project about 1″ beyond the hand. The handle is gripped between the first two fingers and thumb, so that the ring and little fingers close in on the handle very lightly. If the handle is gripped so tightly as to jar the hand when striking a blow, a sore or blistered hand will be the result.

Striking is done as follows: the arm is held out from the body just enough to give good clearance. The hammer thus held is raised until the hand is above the shoulder and the wrist is tipped backward, which tip gives to the hammer a little additional swing; and the blow is produced with a full arm swing — the forearm unfolds from the elbow, and the hand moves downward at the wrist just as the arm reaches its full downward position. The beginner may find that at first he strikes wildly, but with short practise he can deliver a true and very effective blow.

When light blows for finishing a piece are needed, the handle is gripped about the middle between the thumb and four fingers, the thumb resting along the top of the handle. The blow is struck with a forearm movement.

Sledge. — The helper, when using the sledge, should stand in front of the anvil directly opposite the smith, and far enough away so that the sledge will just reach the work; this distance depends on the kind of blow struck and the length of the handle and of the helper's arms. There are two ways of striking with a sledge: (*a*) straight down from the shoulder, and (*b*) a full arm swing in a complete circle. This latter is a much more powerful blow, but is to be used with caution by the beginner. When using the straight downward blow, one grasps the sledge with the right hand at the end of the handle, swings it from the floor to the height of the waist, grasps the handle at about the middle with the left hand, lifts with both hands until the left is over the left shoulder (Fig. 83),[1] and strikes straight downward onto the work, — without changing the hands, barely letting the right hand slip towards the end.

When striking a swinging blow one grasps with both hands close to the end of the handle, and the sledge is placed on the work where the blow is desired. The helper then steps back till his arms are at full length (Fig. 84). The feet are placed at about right angles but the heels should be fully 12″ apart. In this position one can swing the sledge through a circle and down upon his work, producing a heavy blow.

DRAWING DOWN

Drawing down or drawing out consists in lengthening the stock by reducing the area of the cross section, either throughout its entire length or a portion. The cross section may be kept the same shape or it may be changed, as from round to square or from square to octagonal. Small stock usually is worked on the face of the anvil. When the iron is to be expanded, both in length and breadth, the flat face of the hammer is used, but when it is to be stretched in one way only, the pene of the hammer is used, and the work is struck with the pene at right angles to the direction in which extension is desired. This action is that of a dull wedge forcing the metal in the desired direction. The straight pene should be used for widening the stock as shown

[1] It is natural for some people to take hold of the handle with the hands reversed from the position just described. For such persons the description will answer if they substitute the words *right hand* for *left hand*, in each case.

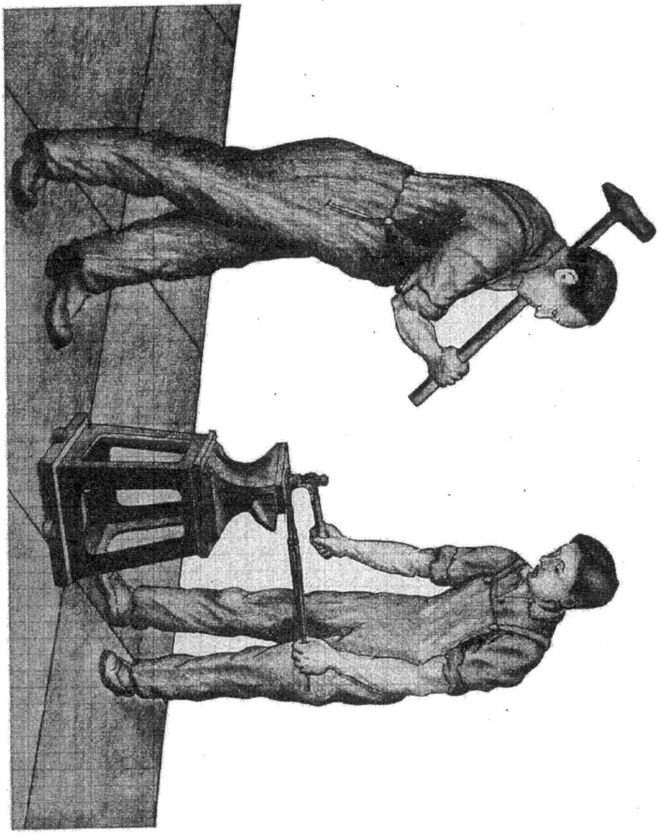

Fig. 83

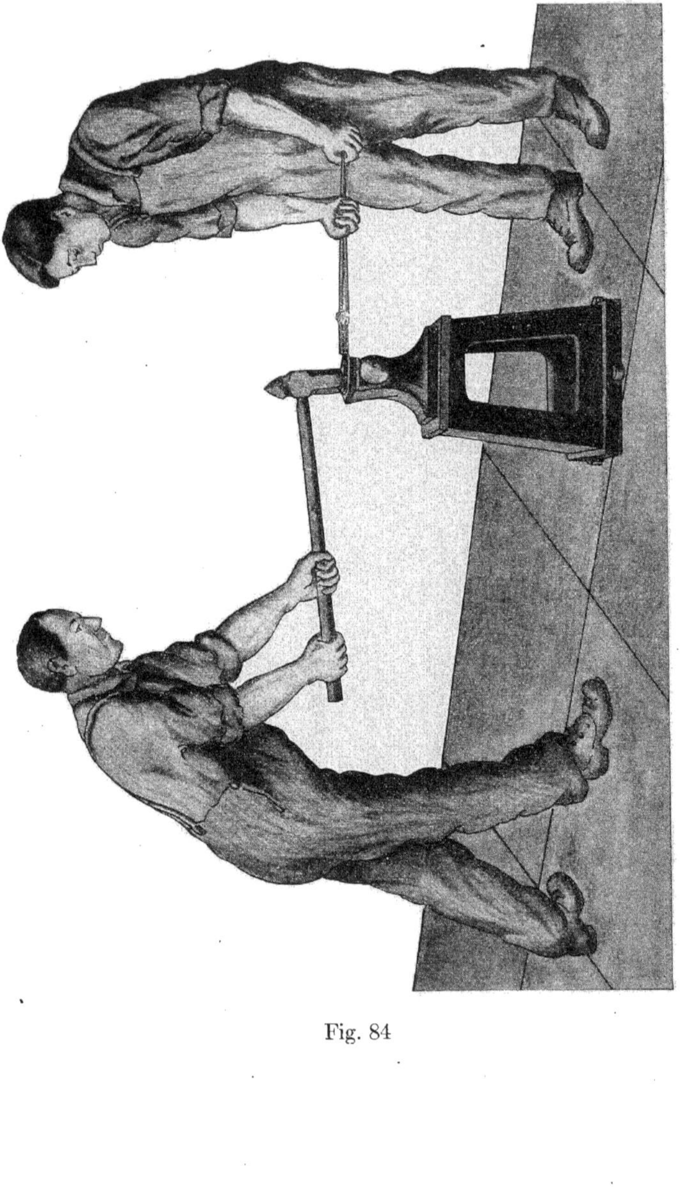

Fig. 84

in Fig. 85; and the cross pene for lengthening it as in Fig. 86. The face of many anvils is crowned and when this is the case the stock can be lengthened by holding as in Fig. 86 and by striking either with hammer-face or pene. When the stock is to be widened it should be held as in Fig. 87. Large work that is to be made much longer, but little if any wider, or is to have its section greatly reduced, can be drawn out much more rapidly on the horn (Fig. 88) than on the face of the anvil, since the rounded horn, like the dull wedge, as explained above, forces the metal lengthwise and prevents the work from widening. When the stock that is being drawn down is to have the

Fig. 85

Fig. 86

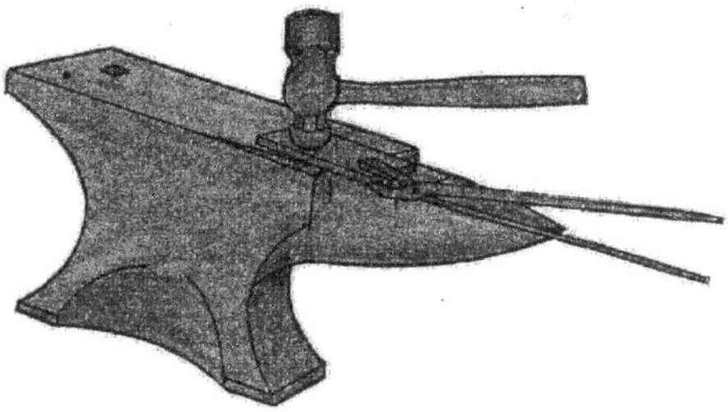

Fig. 87

sides parallel, care must be taken to have the hammer face fall parallel to the anvil.

Drawing Down a Square Bar. — If a bar of square section is to be drawn down square but smaller, the

Fig. 88

bar should be turned accurately a quarter turn at each blow. The art of rolling the stock a quarter turn is not easy, and the first attempts are almost sure to produce a diamond-shaped cross section, but when learned it is remarkable what true work can be produced. If a diamond-shaped section results, it can be brought back to the square by striking as shown in Fig. 89. Rectangular stock is handled in the same way as square stock, after

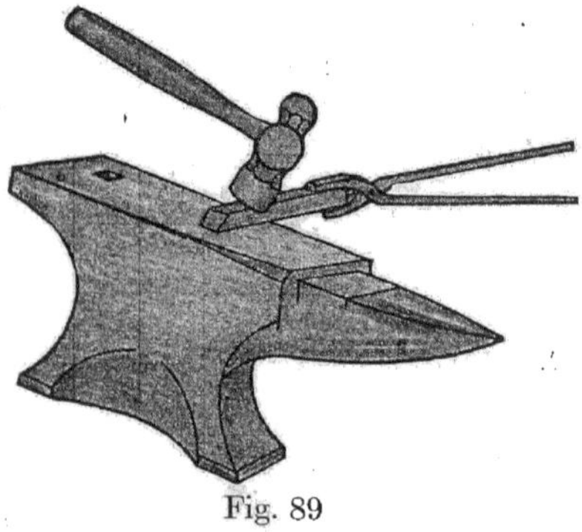

Fig. 89

the sides of the rectangle have been made to bear the proper proportions to each other. Thus if the dimensions of the stock to be drawn down are, say 2 to 1 as in (*a*) Fig. 90, and the desired section is to be 3 to 1 as in (*b*) Fig. 90, then face (*c*) should be struck to lengthen it and shorten face (*d*) until (*e*) is 3 times (*f*); after that the sides should be kept in this proportion by turning a quarter turn after each blow. Changing from rectangular section to square, or from square to rectangular, should be accomplished in the same way. On square or rectangular stock, only two sides need receive the blows. It is good practise for a beginner to hammer a bar of cold iron and observe the indentations. If he does not bring the hammer down flat the marks will tell quickly.

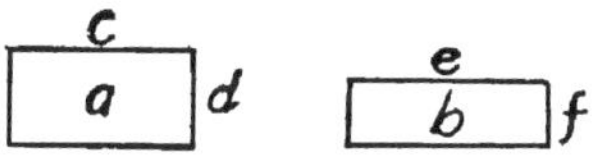

Fig. 90

Drawing Down Round Stock. — Round stock, even if it is to be round when finished, should always be drawn down square to the required size and then rounded with as few blows as possible. If it is attempted to keep the stock round throughout the process, the bar may become cupped at the ends and split through the center, as shown in Fig. 91. The cupping is due to a too rapid working of the outer surface, which causes it to stretch more than the interior. This outer stretching can be carried so far as to cause the outer surface to become entirely loosened from the center of the stock. The cracks are due to a movement of fibers, illustrated by Fig. 92 and explained as follows: a downward blow at (*a*) will tend to distort the circle to an ellipse, as shown by the dotted lines, and when turned slightly for the next blow the sides will roll by each other as at (*c*). This, repeated many times, develops cracks. Fig. 93 illustrates the steps in drawing from round to

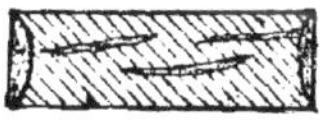
Fig. 91

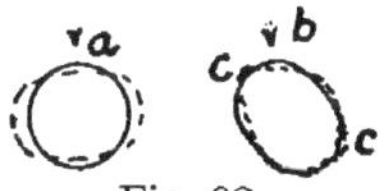

Fig. 92

Fig. 93

round: (*a*) shows the original stock, first the section is reduced to a square, the sides (*b*) of which should measure slightly less than the diameter (*c*) of the piece desired.[1] The four corners are now hammered so as to make the section octagonal. These corners are hammered into eight more sides; and so on, until the section is round.

When drawing down to a round point, one must follow the same course. The end should be drawn to a square pyramid of proper size and length and then rounded by being hammered first to four corners, then to eight, etc. The point always must be kept hot, or it will split. In drawing down the iron should be heated to a bright red and not hammered after it reaches a dull red, except in finishing, when it is hammered with light blows from a dull red to a black.

UPSETTING

Upsetting or jumping up is the reverse operation of drawing down; a piece is shortened in length and the cross-section increased in one dimension or more. Upsetting is a slower and more laborious process than drawing down. There are several methods of upsetting. The proper one to use depends largely upon the shape and size of the work. If the piece is short it is generally stood on end on the anvil and the upper end is struck with a hammer (Fig. 94). Pieces small in

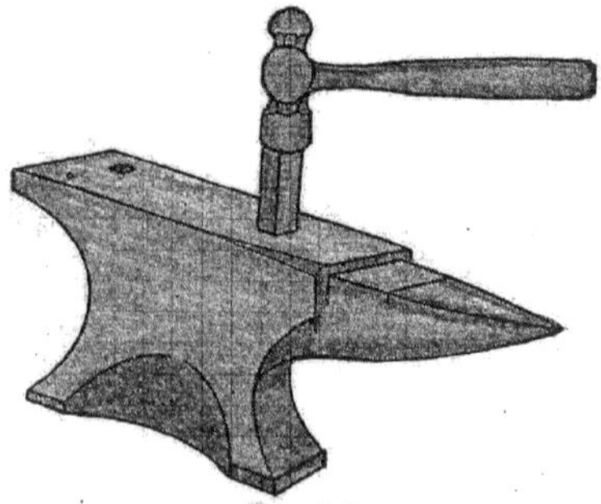

Fig. 94

[1] As the sides bulge out somewhat when the corners are flattened.

cross section relative to their length as a piece of $\frac{3}{8}''$ round 8″ or 12″ long, which is to be upset on the end, should be held flat on the anvil by the tongs, so that the end of the piece to be upset projects over the edge about $\frac{1}{2}''$, where it can be struck a sharp blow with the hammer. By resting the piece on the face of the anvil, the tendency to bend is greatly lessened. The handle of the tongs should be pressed tightly against the left leg, to prevent a backward movement of the work to increase the efficiency of the blows. Pieces of larger size are upset by being bumped repeatedly upon the face of the anvil (Fig. 95) or upon a plate of cast iron set in the floor alongside of the anvil. Another way is to lay the piece upon the anvil face or swing it in a chain, hold the end with both hands and strike the other end with a sledge or a

Fig. 95

Fig. 96

swinging monkey (Fig. 96). Many heats are often required to jump up a moderate mass of metal, and the result is that the dimensions are not very exact. The upset metal, in spite of much care in localizing the heat to the place wanted, is unequal and without sharp shoulders; hence to allow for drawing back to the desired form and size considerably more must be upset than is needed.

Upsetting tends to separate the fibers of the metal. It is therefore necessary that the work be done at a welding heat.

Pieces of any length will tend to bend when being upset and should be straightened as soon as a bend starts, because additional blows will simply bend the stock more and produce no upsetting.

The blows must be heavy enough to work the metal the entire distance that is to be upset, in order that the section may remain uniform throughout. If it is found that the ends are spreading faster than the center of the stock, they can be so cooled that when struck with the hammer they will remain unchanged while the hot middle part is worked. This can be repeated till the piece has become uniform in section.

QUESTIONS FOR REVIEW

Describe the position of the smith at the anvil. How should the hammer be held? How many ways of striking with a sledge are there? Describe each. How is the hammer held for light, finishing blows? What is drawing out? What is upsetting? How should a round piece be drawn down? If in working a square piece it gets diamond shape, how can it be made square again? When and how is the pene of the hammer used in drawing out? What is the action of the pene? When is the horn of the anvil used? Why? What effect has the crown on the anvil face in drawing out? When do we use the flat face of the hammer? Describe drawing down a square bar. A rectangular bar. How do you change from a rectangle to a square and vice versa? How many sides need be hit with the hammer in drawing down a square? How are short pieces upset? Long, heavy ones? Long, light ones? If the ends upset faster than the middle, what can be done to work the middle up? How heavy a blow is needed in upsetting? What is a swinging monkey? Will the metal upset evenly?

CHAPTER VI

BENDING AND TWISTING

Bending and twisting are very important, but comparatively simple.

Curves and rings of light section are easily bent over the horn or round edge of the anvil, or around a suitable mandrel. With heavy sections bending blocks are necessary. It is easier to bend bar iron flatwise than edgewise. Along the center of the bar (*a–a*) (Fig. 97) or the neutral axis, as it is called, the bending is performed without stretching or shortening the fibers, but on the convex side of this neutral axis all the fibers are stretched. The greater the distance is from (*a–a*), the greater the extension. Again, all the fibers on the concave side are shortened or compressed and the greater the distance, the greater the compression; therefore, the greater the distance (*b*) the more the work that must be done in the extension and compression. Also the metal tends to wrinkle or buckle and must be kept straight by hammering.

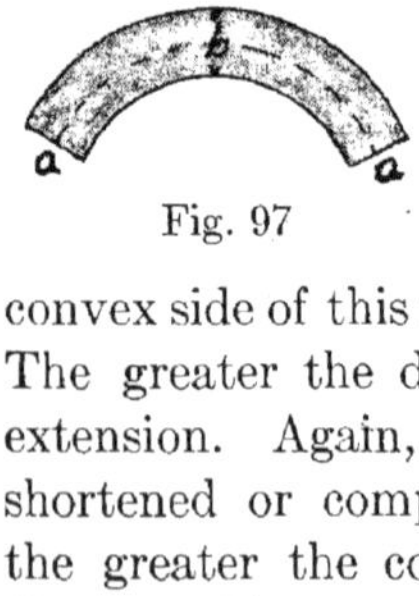

Fig. 97

Flat Bend.—Bending a piece of iron the flat way to some angle is the most simple case of bending. Suppose a piece of rectangular iron is to be bent to a right angle, the corner (*a*) (Fig. 99) to be left rounded. The iron is heated to a bright red heat at the place where the bend is to be made, rested on the face of the anvil with the heated place over the round edge (Fig. 98), and the projecting edge hit with the hammer at (*b*) and (*c*) (Figs. 98–99), till the piece is brought to the desired angle as shown by the dotted lines.[1] It is then trued up on the face of the anvil. The piece can often be bent more easily if a sledge is held as shown in Fig. 98.

Fig. 98

Bending to "U" is done by heating the piece at the place where bending is desired and the portion that will be taken up in the bend, placing the piece with the center of the heated portion over that part of the horn where the diameter is about the same as the diameter of the bend desired, and striking the free end until it is bent to a right angle as shown in Fig. 101. End (*a*) is grasped in the tongs and (*b*) bent in a similar way. The piece is next placed on the face of the anvil and struck at (*c*). If one end should be slightly longer than the other, stand the longer end on

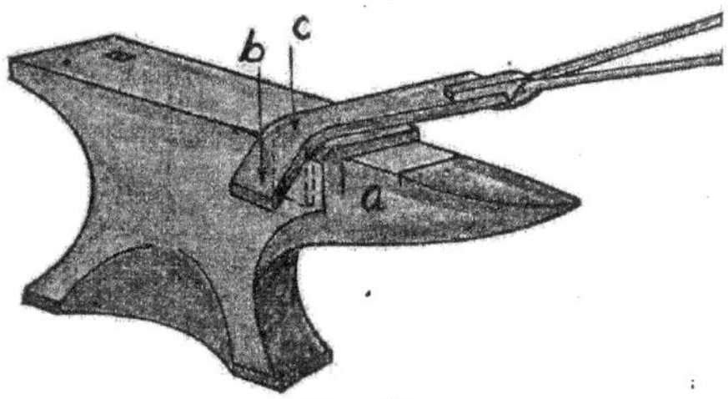

Fig. 99

[1] After the piece has been bent to an angle of 110° to 120° it is often easier to finish the bend with less danger of injuring the piece by holding and striking, as shown at (*a*) in Fig. 100. If one leg is a little long, it can be shortened by making the bend sharper as shown at (*b*) in Fig. 100.

the face of the anvil as at (*a*) in Fig. 102 and strike as indicated by the arrow.

Ring Bending. — After the proper amount of stock has been cut off,[1] about half its length is heated to a dark red and placed over the horn of the anvil, as in Fig. 102, and bent down by being struck in the direction of the arrow. The bending is continued by advancing the piece across the horn and striking as before. This process is continued until about one-half of the piece is bent; then the other end is heated and bent similarly. When bending a piece to form a ring or similar shape, never strike directly over the horn but a little in advance of the support, otherwise the stock will be marred or hammered out of shape. If the ring is to be welded, the ends will have to be scarfed as explained in the chapter on Welding. If it is to be unwelded but with ends flush as (*a*) in Fig. 103, each end will have to be beveled as at (*b*) in Fig. 103, by an amount determined by practice.

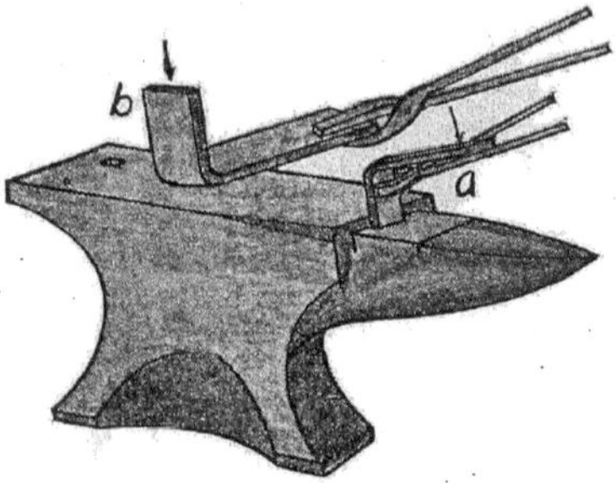

Fig. 100

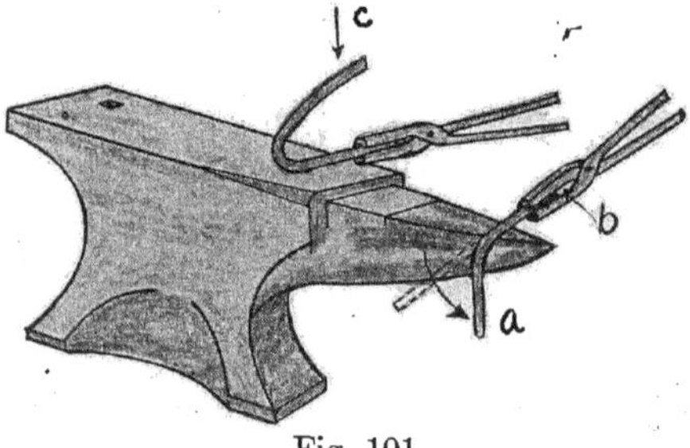

Fig. 101

Eye Bending. — If an eye like (*a*) (Fig. 104) is wanted, the length of the stock necessary to form the eye is de-

[1] See calculations for ring, Chapter XVI.

termined. Then the piece is heated at a point as described above and shown at (*b*) (Fig. 104). The piece is next heated at the end and placed over the horn and

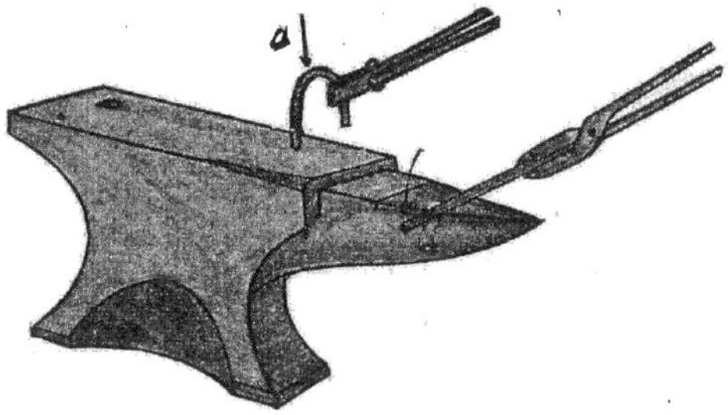

Fig. 102

bent, as shown at (*c*) and (*d*), in much the same way as was the ring (Fig. 102). The eye is then closed by bringing down the end, to form as shown at (*a*), and by holding and striking as at (*a*) and (*b*) (Fig. 105). To close the eye in properly it may be necessary to place the piece in positions (*a*) and (*b*) two or three times and to make the center line of the stem pass through the center of the eye (*a*) (Fig 104), and likewise to round the eye it may be necessary to place the piece over the horn as shown at (*c*) (Fig. 105).

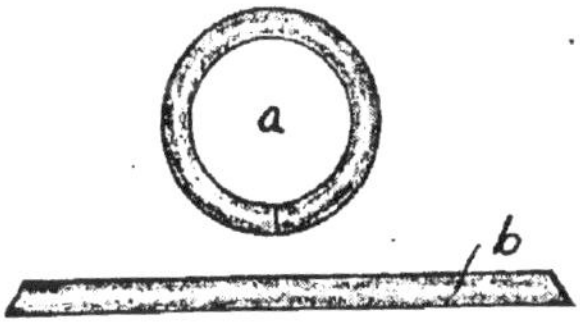

Fig. 103

Hook. — A hook is bent somewhat similarly. It is bent at right angles (*b*) (Fig. 104), placed over the horn (*a*) (Fig. 106), and with further blows carried around to the position indicated by

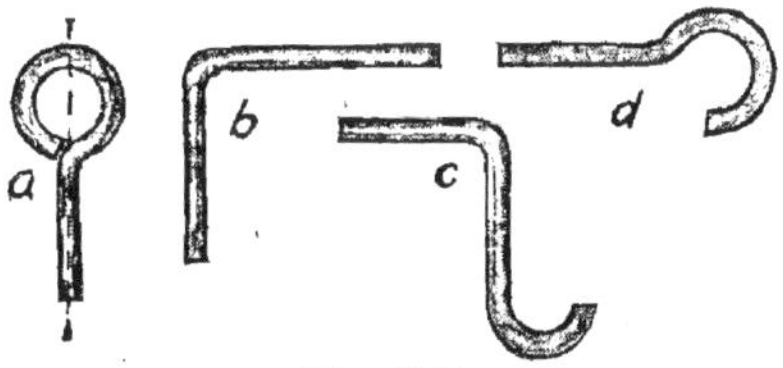

Fig. 104

the dotted line. The bend in the end of the hook is produced over the edge of the anvil (*b*) (Fig. 106).

Edge Bend with Square Forged Corner. — This is an especially difficult piece to make without having cracks form as shown at (*a*) (Fig. 107). The stock is fullered[1] and drawn down as shown at (*b*) in Fig. 107,

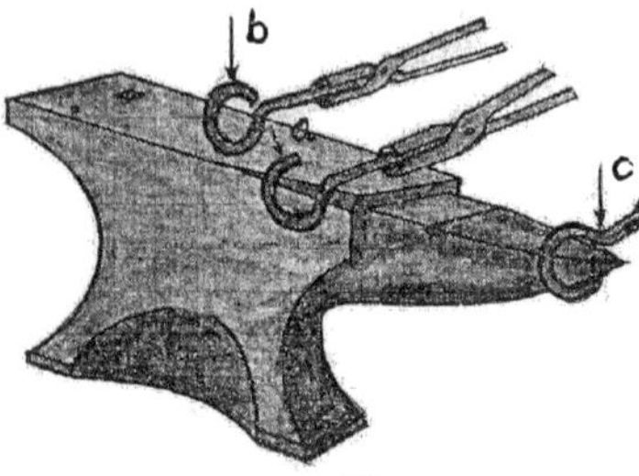

Fig. 105

the projection shown at (*c*) left where the corner is to be, is heated almost to a white heat, and bent over the round part of the anvil face to an angle of 110° or 120° (Fig. 108). It will be well to have a helper hold a

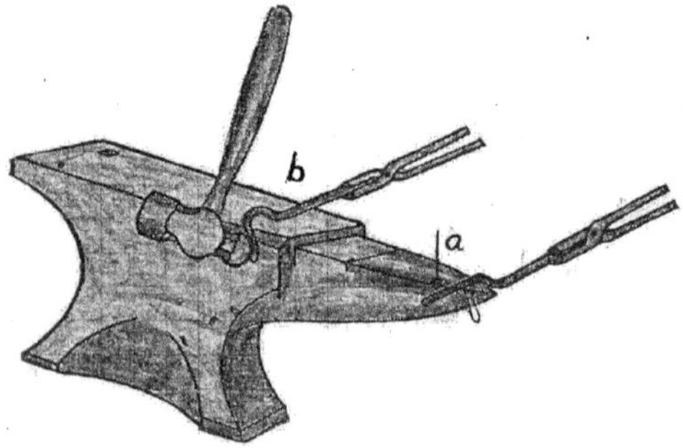

Fig. 106

sledge on the piece so that the edge of the sledge is directly over the edge of the anvil. The bending can be done by placing the work in the vise, but this is not recommended as the stock is likely to be cut. The stock is reheated at the bend by being placed in the fire

[1] See Chapter VIII for description of fullering.

as in Fig. 109, and the corner is worked up square by a series of blows given as follows: The work is placed over the anvil, as in Fig. 108, but so that the inside of

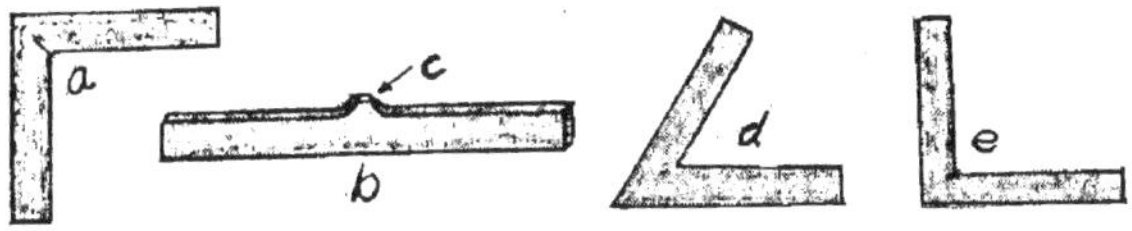

Fig. 107

the bend is not allowed to touch the corner of the anvil; struck as indicated by the arrow; turned, end for end, and struck again; placed on the face of the anvil as (*a*) in Fig. 108; and struck on the ends. These blows are to be repeated, with occasionally a blow to bring the piece nearer to a right angle, until a sharp corner is made on the outside and the piece is nearly at a right angle. The corner can now be finished by standing the piece on end and striking as indicated by the arrow in Fig. 110 and finally closing in to the exact right angle. In all this work the corner must be kept at a good heat; the angle must be more than 90° at all times with a small fillet on the inside. Never let it get into the position shown at (*d*) (Fig. 107) or the metal will lap over and produce a crack as at (*a*), or a cold shut as at (*e*). The piece should be finished all over with a flatter. A bend with a square corner as shown at (*a*) (Fig. 107) can be forged without being fullered and drawn down. All other operations

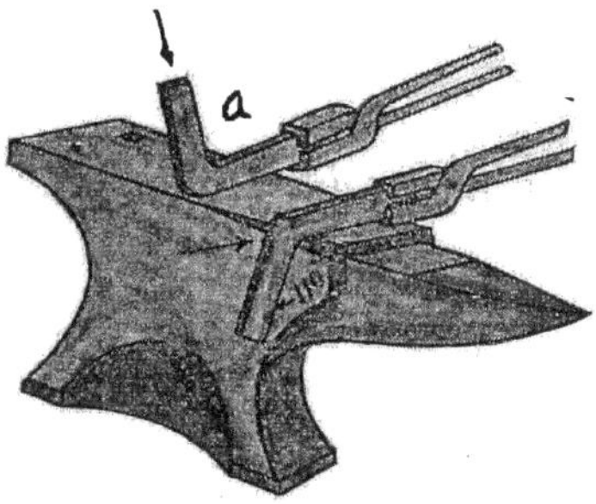

Fig. 108

Fig. 109

are the same as described above; but the process is more difficult.

Bending Plates. — In bending large work to various outlines many devices are used, such as cast iron templates, or bending blocks. Their design and contruction are a matter of cost. When only a few simple pieces are to be bent the cost of even a simple block would be prohibitive, but when many pieces are wanted all alike, elaborate blocks may be made. Fig. 111 shows a block for bending flat bars. It can be made for any radius or even for differently shaped curves, and the working edge (*a*) can be made with a *T* or *L* section; so that "shapes" also can be bent. The block consists of a heavy cast-iron plate of the desired shape with a means suitable for fastening it to a bench or leveling block. The end of the piece to be bent is held in the groove (*a*) by the clips (*b*), which are fastened to the plate by bolts or pins in some of the holes and by the pin (*c*), slipped through the slots in (*b*). As the piece is bent more clips are added, which process holds it to the block until the required amount of the piece is bent to the required shape. Fig. 112 shows a bending block for general work, fitted with a screw. Upon this block bars can be straightened or bent at most any shape. The block is pierced with numerous circular and slotted

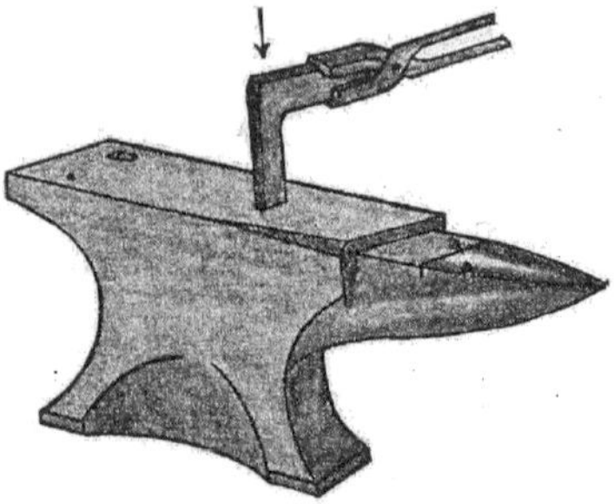

Fig. 110

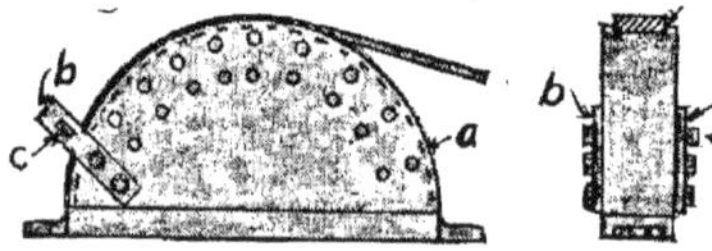

Fig. 111

holes, which receive pins to form the necessary supports for the piece in bending. At (*a*) is a block securely fastened to the plate, in which is a thread to move the screw (*b*) in or out to put pressure on the work. In operation the plate is very simple. Pins are placed in the holes, the piece is forced between them, and is bent to the desired shape. The screw is useful in straightening stock and is used as follows: Pins are placed in holes (*c*), the piece to be straightened placed so that it bears against these pins, with the bowed side to the screw, and pressure applied through the screw till the piece is straight. Numerous devices for bending special forms can be made and attached to this form of plate.

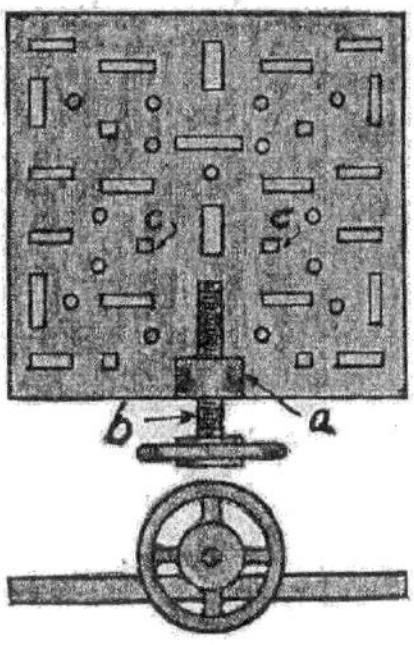

Fig. 112

Twisting. — To make the twist as shown in Fig. 113, lay off the portion to be twisted and lightly center punch at each end. If the stock is light enough to be twisted without heating, place it in a vise with one punch mark just above the jaws, the rest of the stock extending above. Then with a wrench that fits tightly, grasp the stock just above the second punch mark and give the piece the desired number of turns. If the left hand is used to back up the wrench it aids in keeping the stock straight. When the stock is large and needs heating the procedure is the same. The stock must be heated uniformly and the work performed quickly, or the cold vise and wrench will extract heat from the stock near the place where it is held, and results in an uneven twist, since the hotter portion will turn more easily, giving a shorter

Fig. 113

twist. It is also hard to keep the piece from bending as the wrench cannot always be backed up by the left hand owing to the heat. When the piece comes out crooked it can be straightened without marring by heating it to a dull red, and hammering it between two hardwood boards.

QUESTIONS FOR REVIEW

How are curves and rings bent? What should be used for heavy stock? What is the neutral axis? How is the metal disposed of that is bulged out by compressing in making a bend? Describe how to make a flat bend. Describe bending to a U. How is a ring bent? Why should the piece never be struck directly over the horn? Why is the piece first bent to a right angle in making an eye bend? In the edge bend, why is it easier to make the piece by the first method described? Why must the inside corner be kept away from the edge of the anvil after it is first bent? Why in making the edge bend, must the piece always be kept at an angle greater than a right angle till the corner has been forged sharp? What are bending plates? When is it profitable to use them? Describe how they are used. Describe the operation of twisting. Why must one work fast when twisting a piece of hot iron? Why can a piece of cold iron be twisted more evenly than hot?

CHAPTER VII

SPLITTING, PUNCHING, AND RIVETING

SPLITTING

SPLITTING is done by hand with an ordinary hand chisel, with a hack saw, and by power with a slitting shear or slitting saw. The methods of splitting by hand are the only ones which are described in this work.

Splitting with the Hot Chisel. — The heated piece is placed on the cutting block (*a*) (Fig. 114), the chisel held by the smith on the place where the cut is to be made, and struck by the helper with the sledge till the stock is cut through. Unless the stock is very thick, it is well to cut through all the way from one side. The piece can be held on the hardie and the chisel on the stock directly above (*b*) (Fig. 114), but this is more difficult. Before stock is split, it is necessary to punch or to drill a small hole at the place that will be the end of the split. In Fig. 115 (*a*) shows the hole and the dotted line is where the piece is to be split. Light pieces are most easily split in a vise with

Fig. 114

Fig. 115

an ordinary cold chisel, as shown in Fig. 116. The piece is set in the vise, so that the place selected for the split is flush with the top of the jaws, the cutting edge of the chisel being held on the top of the vise, and driven with the hammer against the work, so that the back jaw of the vise and the chisel act as shear blades to cut or split the stock. The method of splitting with a hack saw is the same as that of ripping a board held in a vise. With reference to Fig. 117, no further explanation is necessary.

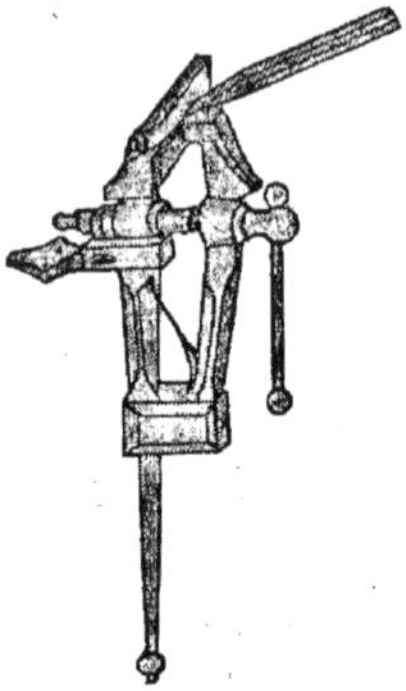

Fig. 116

PUNCHING

Punching is an important operation in the forge shop, though the introduction of the drill press has lessened its use somewhat. There are many cases where the punch is desirable, as in punching eyes for hammer handles. There are both hand and power punches which will make holes of almost any section.

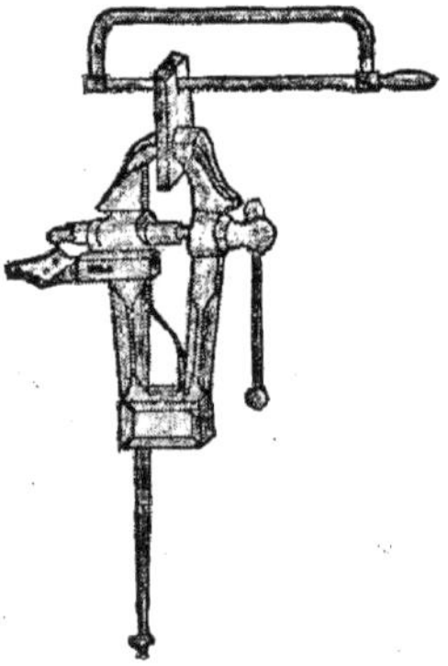

Fig. 117

Power Punch. — The operation of the power punch is very simple. The piece to be punched is held under the punch and a lever pressed down with the foot. This throws a clutch which starts the gears; the punch descends and penetrates the piece.

Hand punches are commonly of two kinds. That shown in Fig. 118 is held in the left hand of the smith and driven into the stock with a hand hammer. That shown in Fig. 55 is held by the handle in the right hand, similarly as the chisel in Fig. 114, and is driven into the work, with the sledge by a helper.

The punch (Fig. 118) is used for small holes in thin iron. It is made from octagonal steel, eight or ten inches long. The end is forged tapering to

Fig. 118

the same shape but slightly smaller on the end than the desired hole. The end should be perfectly flat and at right angle to the center line, as in Fig. 118. As the operation of punching is the same with either punch, a description of the use of one will answer. The iron should be heated to a bright red or almost a white heat, and laid flat on the face of the anvil, the punch placed in position and held to the work firmly with the hand, and driven a little over half through (*a*) (Fig. 119).

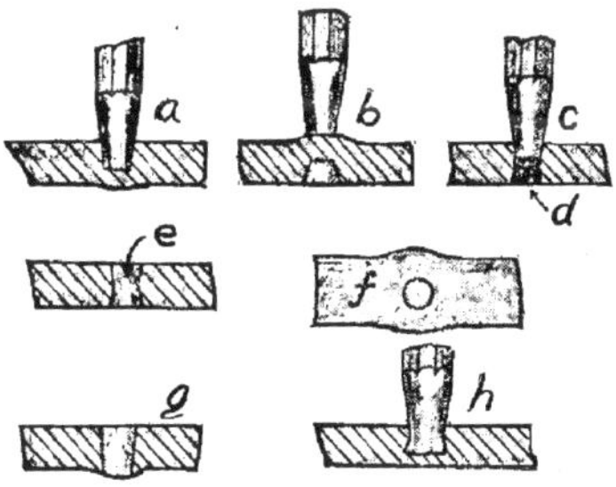

Fig. 119

This compresses the metal underneath the punch and either raises a slight bulge on the under side of the bar, or makes a darkened spot the shape of the end of the punch. The piece is now turned over and the punch placed on this bulged or dark spot (*b*) (Fig. 119). The punch is again driven about halfway through (*c*), and then the work is moved over the punch hole in the end of the anvil and the punch driven through. The slug is thus driven out and the hole (*d*) results. The piece now has a hole through it, as shown at (*e*), but if the stock is narrow it likely will be bulged as shown at (*f*). This bulge must be hammered back to the original width of the bar, without making the hole elliptical by leaving the punch in the hole. In this operation, the punch must be changed from one side of the hole to the other, or the hole will be tapered the same as the punch. The punch must not be driven all the way through from one side, or the result will be a tapered hole and the stock will be bulged on the under side as shown at (*g*), which will make the work look rough, no matter how much one tries to remedy the defect by hammering the bulge back. On thick work, and especially on steel, the punch can be prevented from sticking by placing a little coal dust in the hole just after it is started. The punch must be dipped in water occasionally to cool the end, or it will soften and bulge and thus rivet itself in the hole as shown at (*h*), making it almost impossible to remove it.

RIVETING

Riveting is the joining together of two or more pieces of metal by another piece of metal called a rivet,— which is inserted through holes in the pieces to be joined; after which the ends of the rivet are flattened down (as shown in Fig. 120) to prevent its coming out. Rivets are designated by the shape of their heads as (Fig. 121); (*a*) round head, (*b*) conical head, (*c*) countersink, (*d*) pan head, etc. The conical head rivet has the least cross-sectional area to resist the strain, and the pan head the most. The joints are spoken of as lap or butt joints and as single, double, etc., according to the way the joints are made and to the number of rows of rivets. Fig. 122 shows a butt and Fig. 123 a lap joint. The distance between the centers of two rivets adjacent in a row is called the pitch.

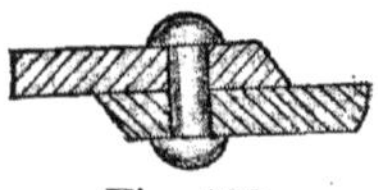

Fig. 120

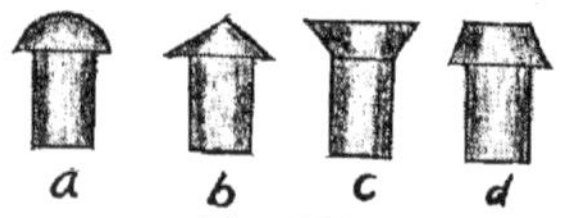

Fig. 121

The way in which a rivet is driven depends upon the purpose of the rivet. If it is to make a tight joint or seam, the pieces are brought together into their proper position. The rivet is heated to a full red heat, passed through the hole previously punched or drilled, held in place by a dolly-bar[1] or rested on the anvil according to the nature of the work, and then driven with heavy blows to upset the stem and fill the hole. The head is afterward rounded to the shape desired. The quicker the riveting is done the more heat is left in it and hence the greater the amount of contraction, after the riveting is finished, to draw the plates together. If the rivet is to fill the irregularities of a punched hole, it should, when heated, be as good a fit in the hole as possible.

Fig. 122

Fig. 123

If the rivet is to hold two pieces together like a pair of tongs, where there must be movement, the rivet is struck light blows which spread out the end to the desired shape but does not upset the stem. The heads of rivets are usually finished in a cupping tool (Fig. 57). After the head has been hammered into shape, usually with the pene of the hammer, the cupping tool is placed over the rivet head and struck a few blows with the face of the hammer. Riveting on structural and boiler work is usually done with a pneumatic riveter.

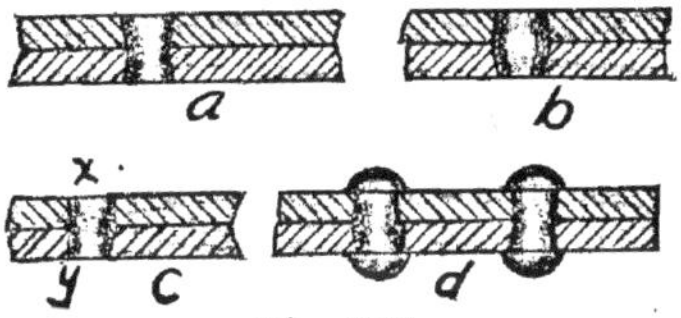

Fig. 124

The effect which the taper in a punched hole has

[1] A dolly-bar is a heavy piece of iron 18″ to 2′ long, which is held against a rivet to act as an anvil so as to upset the rivet and form the head.

upon a riveted joint depends upon the manner in which the plates are brought together. In Fig. 124 at (*a*) are two holes with the large end of the taper on the outside, and the effect is the same as a slight countersink, causing the rivet to draw the plates together. It is evident that this desirable method will make a tight joint. It has two drawbacks, however: punching from opposite sides is difficult; in making repairs it is hard to remove the old rivet without injury to the plates. At (*b*) are shown plates placed together with the large diameter of the holes inside. This is bad, since the tendency in driving the rivet will be to spread the plates apart. This rivet would also be hard to remove for repairs. (*c*) shows the plates both punched from one side. This will make it hard to drive the rivet and get a tight joint and the rivet will also be hard to remove from side (*x*). But it can be easily removed from side (*y*), if it is accessible. At (*d*) are shown cases where the rivets do not come fair. In these cases the rivets bend in driving and the result is that they do not fill the hole.

QUESTIONS FOR REVIEW

What are the ordinary ways of splitting stock? Describe a hot chisel. How is it used? Why is the small hole needed at the end of the place where the split is to be? How are light pieces split in the vise? Why is punching better on some classes of work than drilling? Describe the method of punching a hole through a piece of iron. Why must the hole be punched from both sides? Why must the punch be left in the hole when the bulged sides are hammered back? Why must the punch be changed from one side to the other in hammering back the bulged sides? Why must the punch be dipped in water often while it is used? How is the punch prevented from sticking in deep holes? What will happen if the end of the punch is allowed to get soft? What is riveting? What is a rivet? How are rivets named? Give the names. What is meant by pitch, double riveting, butt and lap joints? What is a dolly-bar? Why should riveting be done hot when a tight joint is to be made? Why are light hammer blows used when a pair of tongs are riveted? Why is a cupping tool used? What is the effect if the large diameters of punched holes are on the outsides? on the insides? one inside and one outside? If the holes do not come fair?

CHAPTER VIII

THE USES OF BLACKSMITHS' TOOLS

FULLERING

THE fullers (Figs. 66–67) are used to change the direction of the fibers of iron (without cutting) where it is to be reduced in section, or to widen stock. Fig. 125 (*a*) shows the use of the top fuller alone in making a depression in a piece of tool steel preparatory to making a lathe tool. The steel is held on the face of the anvil and the fuller is held on the steel at the proper place, while the helper strikes the fuller with a sledge which causes it to sink into the steel. Fig. 125 (*b*) shows the use of both the top and bottom fuller in position for forming shoulders. In this case the ends are to remain at their original size and the part (*c*) between the fuller marks is to be reduced and rounded. The operation is as follows: place the bottom fuller in the hardie hole and heat the iron to a good red heat; hold it on the fuller at the place where the shoulder is to be (the top fuller is now held on the iron directly over and parallel to the bottom fuller)[1] and strike the

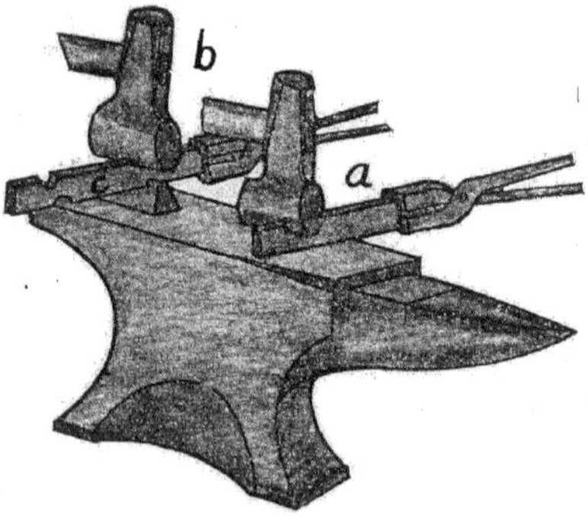

Fig. 125

[1] Caution: Hold the top fuller so that it will be parallel in both planes with the bottom one.

top fuller with a sledge. The stock must be turned repeatedly from one side to the other in order to have the two fuller marks the same depths: the two fullers are scarcely ever of exactly the same radius, and the sharper one cuts the faster. By the turning of the piece each side is brought for an equal time into contact with the sharper fuller. Fig. 126 shows another use of the fuller; namely, that of spreading a piece of iron or steel. For this use the fuller is held in an inclined position and driven with a sledge, as indicated by the arrows, which forces the metal as shown in the figure. Fig. 126 illustrates the use of the fuller for both shouldering and spreading. When forging a piece to shape a block of proper length is taken and first fullered with the top fuller, as shown at (*a*) in Fig. 125, the piece results as shown in Fig. 127. The fuller is next applied as shown in Fig. 126.

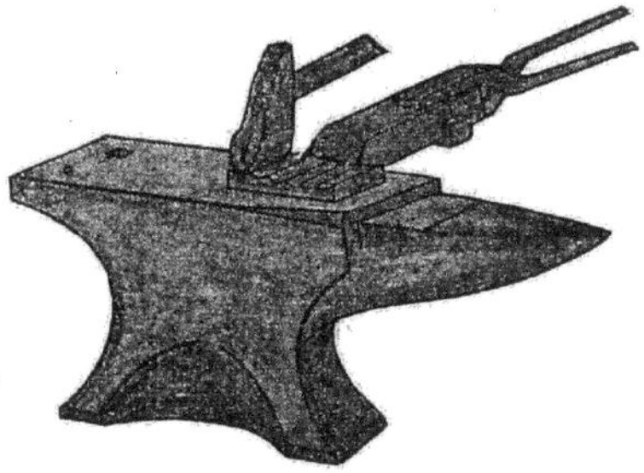
Fig. 126

Fig. 127

SWAGES

Swages shown in Figs. 61 and 62 are used for finishing work. Swages for round work may be semicircular, as (*a*) Fig. 128, or V-shaped, as (*b*) in Fig. 128. The semicircular swage makes the neater, more nearly circular job, but is more apt to forge the work hollow as explained under drawing out, page 63, Chapter V. For this reason the circular form is used on small work and for finishing; while on large work

Fig. 128

and under the power hammer, the V-form is used to bring the work down to size, which is then finished in the semicircular swage. The spring swage (Fig. 63) is used on light work when the hand hammer alone is used, as the top swage guides itself so that the smith, holding the work in one hand and the hammer in the other, is enabled to use the swage without the aid of a helper. The spring swage is also used under the power hammer, for the same reason. The holes of circular swages are always made to a larger radius than the radius of the work; also they are never half-circles. On small sizes they are about two-thirds of a semicircle and on large sizes less. This is necessary so that the hole between the two swages when placed together will be oval, and thus prevents the swages from wedging upon the work.

The V-shaped swages are, as stated before, for drawing-down work, while the round are for finishing, and should not be used except for the last finishing touches. A novice will find the swage a rather hard tool to use. He usually can get better work with his hand hammer.

Swage-blocks (Fig. 31) should be constructed so that the holes (*a*) passing through them are true circles or squares and the sides parallel the full length of the hole. The recesses (*b*) should be less than semicircles, as in the case of hand swages. The slots at (*c*) should be parallel throughout their length but should taper in depth, to be narrowest at the bottom.

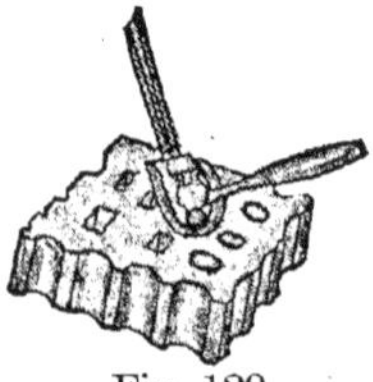

Fig. 129

Operations. — The bottom swage is placed in the hardie hole, the work laid in the groove, the top swage placed on the work directly over the bottom swage and struck light rapid blows by the helper using the sledge, while the smith moves the piece backwards and forwards and revolves it at each blow. Care must be taken that the grooves always remain parallel, and that the work never is placed in swages where the radius of the grooves is not larger than the work.

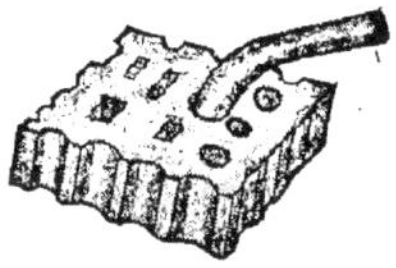
Fig. 130

The grooves on the edges of the swage-blocks are used as bottom swages. The holes are for various purposes; such as to true up stock by passing it through them and shaping and turning edges as in Fig. 129, and to bend pipe (Fig. 130).

FLATTER

The Flatter (Fig. 59) is used to finish flat surfaces, just as swages are used to finish round surfaces. The anvil takes the place of the bottom swage as a bottom flatter. Its use and that of the set hammer are so nearly the same, that the description to be given for the set hammer will answer for both.

SET HAMMER

The Set Hammer (Fig. 58) is used to set down square shoulders on work similarly to the way the fuller is used for rounded corners. It is also used to forge pieces that cannot be reached with the hand hammer, and often in place of a flatter.

Fig. 131 shows the use of the set hammer on setting down stock so as to leave the stem-reinforcing shoulder. The stock was originally all of the same thickness. The set hammer is placed on the piece so that its edge is in line with the side of the stem, and given a blow with the sledge. It is then moved to the other side of the stem and given another blow. The piece is then held at the place indicated by the arrow (*a*) and the end of the stem is set down. At (*b*) Fig. 131 is shown the way the head is spread and narrowed by use of the set hammer.

These two operations must be alternated till the head is as desired.

Heading Tool. — Heading tools are of two types, the hand heading tool (Fig. 72) and the floor heading tool (Fig. 132). They are used to shape and finish heads on bolts and similar articles, or collars on shafts.

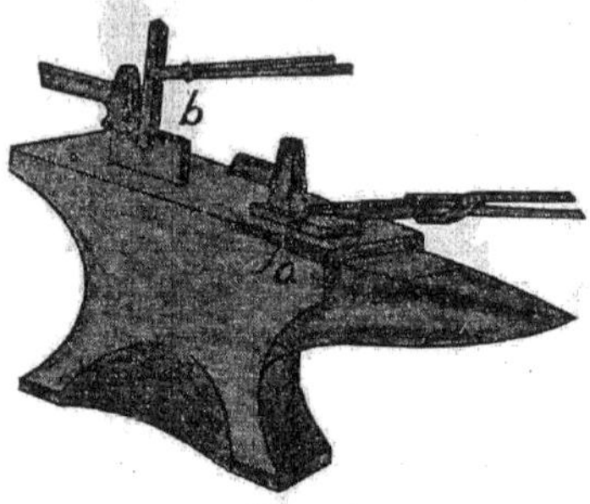

Fig. 131

The hand tool is used as follows: a sufficient amount of metal is jumped up at the desired position; say, at the end of the stem as in making a bolt; heated to a red or almost white heat; placed in the tool (Fig. 133); and struck a few good blows with a hand hammer or sledge (according to the size of the work), to flatten out the head (*a*) and work the under side to a flat face with a sharp corner (*b*). After one or two blows are struck the stock should be removed from the tool and examined to see whether the stem and the head are concentric. If they are not, the head can be shifted by striking it in the direction it should be moved as indicated by the arrow at (*c*). Then the piece is removed from the tool and brought to shape on the face of the anvil (Fig. 134). The head as it leaves the heading tool is round, as shown at (*a*); it is struck a good hard blow, indicated by the arrow, to flatten its two faces, as shown at (*b*). It is then rolled over to the position (*c*) and struck, as shown by the arrow. This leaves it square, as at (*d*). When it

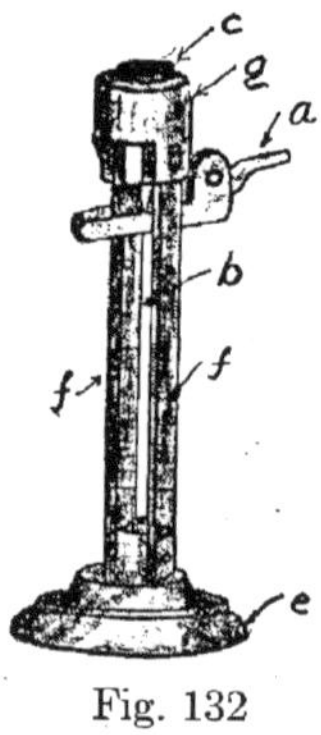

Fig. 132

is being shaped (Fig. 134) the head probably will become too thick and the other dimensions too small. In this case the piece must be placed in the tool and hammered to the correct dimensions. The blows that shape the head must be heavy enough to work the entire head or it will become cupped (Fig. 135). When the tool is placed over the hardie hole, care must be taken that the stem of the piece which is being worked does not touch the anvil as at (*d*) (Fig. 133), or the stem will be injured.

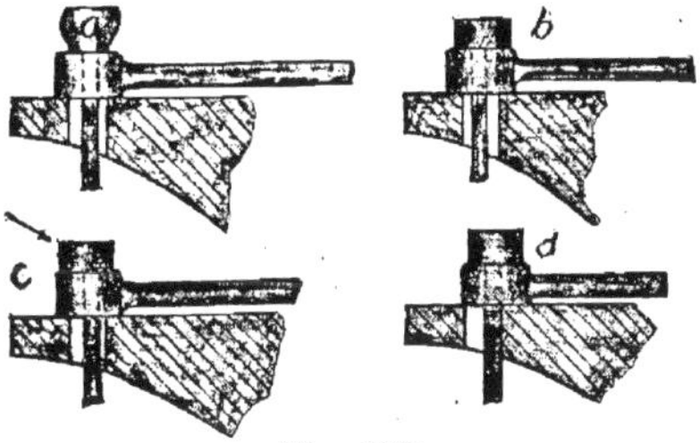

Fig. 133

The Floor Heading Tool (Fig. 132) consists of a base (*e*) and a casting (*g*) connected by two uprights (*f*).

The upper casting is constructed to receive a hardened steel die (*c*) which can be changed for each diameter of stem desired and thus the tool is adapted to a wide range of work. At (*a*) is a lever on which the end of the stem rests when the head is upset. It is also used for forcing the piece out of the tool.

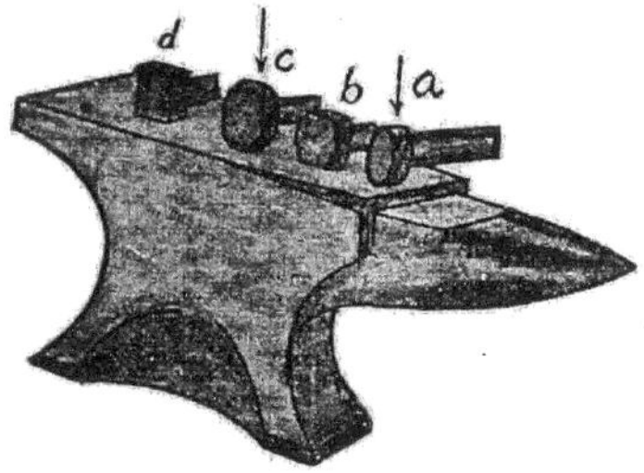

Fig. 134

This tool is a combination upsetting and heading tool. The piece (*a*), which can be set at any desired position in the space (*b*), is placed so that the distance from the top of the die (*c*) to the top face of (*a*) is the length of the required bolt. The stock must be cut just the right length (determined by calculations). The portion that will project above the tool is heated to a white heat. The stock is placed in the tool, the heated portion up, and is given heavy blows with hammer or sledge, which will upset and head

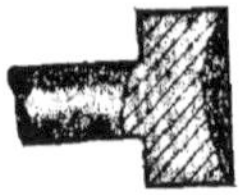

Fig. 135

the piece at the same time. After the second or third blow it should be examined to see if the head is concentric with the stem, and if not, it should be made so, as described above. Care must be taken to keep the stem below the die straight, and not to let the end that rests on (*a*) become burred or it will become too large to pass through the die. The piece can be removed by striking the lever (*a*).

QUESTIONS FOR REVIEW

Describe a fuller and tell what it is used for. What is the action of a fuller in spreading stock? Describe a swage and tell its use. What are circular swages used for? What are V-shaped ones for? Why are circular swages made larger than the work to be finished in them? Why are they always less than a half-circle? What is a spring swage and what is its use? Describe swage-blocks and tell their use. Describe the operation of using a swage. What is a flatter? Give some of its uses. What is a set hammer? Give its uses. What is a heading tool? What is it used for? How many kinds of heading tools are there? What is the advantage of the floor heading tool? How can the head be moved in case it is not concentric with the stem.

CHAPTER IX

HAND WELDING

We have learned the process of shaping iron and are prepared to take up the study of joining two or more of these pieces or the ends of a piece by the process of welding.

Welding. — Welding consists of heating to a welding heat (or nearly to the point of fusion) two or more pieces of iron or steel,[1] at the places where the joint is to be made, and uniting them by pressure or by quick sharp hammer blows.

The exact temperature of the welding heat in wrought iron and steel is not known, but when it is reached, wrought iron and steel become pasty, so that they will stick to similarly heated pieces when placed in contact.

Heated beyond this point, the iron will burn, giving off scintillating sparks.[2] When wrought iron has reached the welding heat it has a dazzling white appearance, and when exposed to the air it makes a slight hissing noise and gives off the bright sparks of burning iron. Soft steel becomes grayish white, and tool steel, bright yellow. As stated above, the metal becomes waxlike or semifluid, a condition in which the particles of the separate pieces come into the same close contact as those

[1] The two ends of a single piece bent to form a ring or chain link, etc., are here considered as two pieces.

[2] Often small pieces of iron or scale get into the fire and burn, giving off scintillating sparks, which is misleading to the smith. This can and should be avoided by keeping the fire clean and deep.

of the solid bar, as they are welded or hammered together. Only such metals can be welded as gradually become softer and softer with increase of heat or those that change slowly from the solid to the liquid state; the greater the range of this semifluid temperature, the more easy is the process of welding. Metals that remain hard up to the crumbling or melting point cannot be welded.

The difficulty in welding is to heat the metal properly and to keep it clean and free from scale. It is *absolutely necessary that the fire be kept clean,* and that its depth be great enough to prevent, as far as possible, oxidation of the metal which results in a dirty heat (that is, the metal will have cinders and dirt adhering to it which will prevent the particles of metal from coming in contact so as to join or weld). Depressions that will pocket air between the pieces will have like effect. The pieces that are to be welded are usually "up set," or enlarged at the places where they are to join, to allow for unavoidable drawing out in the making of the weld and in the subsequent hammering (hammer refining) to refine the grain (which is carried on until the pieces are at a low red heat).

Hammer Refining. — The object of this after-hammering is to break up the coarse crystals that are formed by the high temperature, which can be done by hammering or finishing at a dull red temperature, thereby giving the finished piece a fine strong grain. This finishing to produce the fine grain is called "hammer refining." Care in hammer-refining and proper welding will keep the metal at and close to the weld nearly as strong as the orginal bar. But there will nearly always be a point not far from the weld that has been overheated and has not been hammer refined which will remain a weak place in the work.

Fluxes. — The higher the iron is heated the easier it will take up oxygen and form scale. Scale will prevent a weld whether formed in the fire or in the time consumed in taking the pieces from the fire and placing them together. At the temperature of welding scale is formed very rapidly, so a flux is used to prevent its formation or to dissolve it. When irons of different composition are to be welded together, a flux is needed to prevent the metal that reaches its welding temperature first from oxidizing. This flux will melt and cover the parts to be welded, preventing the formation of scale and dissolving that already formed. When the pieces are placed together and hammered this flux is forced out and the pieces are allowed to join. The fluxes generally used are sand, borax, or a mixture of borax and some substance, as sal ammoniac. Most all special welding compounds have borax as their base. Sand makes a good flux for wrought iron but is of little use with steel. It acts by uniting with the oxide to form a fusible silicate. When steel is being welded, borax or a mixture of borax and sal ammoniac is to be preferred. With this flux, iron scale dissolves at a comparatively low temperature and in this fluid condition can be squeezed out of the way; while if the flux were not used, the iron would have to be subjected to a higher heat to melt the scale. With ordinary wrought iron a heat that will melt the scale can easily be reached without the use of a flux, but with certain machine steel and all tool steel a temperature high enough to melt the oxide would burn the steel.

Borax contains water which should be driven off. This dehydrated borax forms what is called borax glass, which when pulverized makes an excellent flux. Sal ammoniac added to borax in the proportion of 1 part sal ammoniac to 4 parts of borax will act better than borax alone, especially on tool steel. The flux acts merely to protect the surfaces from oxidation and to dissolve the scale, therefore it must not be imagined that it acts in any way like a cement or that welds cannot be made without its use, for with due care to prevent the dirt and oxide they can.

The Procedure. — The process of welding is very simple, but it is one to which too much care cannot be given. As previously stated the fire must be clean and deep, for unless the fire is exactly right the pieces cannot be heated sufficiently or kept free from dirt and scale. If the metal is insufficiently heated or is dirty (covered with scale or slag) no amount of hammering will cause the pieces to join. Again, if the iron is heated too hot, it will burn and become useless, for burned iron will not weld. In welding, everything must be in readiness; the anvil must be clean and nothing in the way, and the hammer laid in a convenient place and in correct position to deliver the blow. The tongs should fit the work tightly and be so held that no changes will be necessary to place the pieces in proper contact. The iron is heated slowly[1] at first until it is at a uniform temperature throughout and then the blast is increased until the welding temperature is reached, then it is taken from the fire and the oxide dissolved off, usually by dipping the metals in a flux and returned to the fire for an instant.

When the pieces are at the proper heat one must work rapidly: the blast is shut off (when power blower is used); the pieces are taken from the fire, given a sharp

[1] If it is heated too rapidly the outside will be at a welding temperature while the interior remains too cold. If it is taken from the fire in this condition it will not weld, because the surface will be cooled below and the welding point by transfer of heat to the interior and by radiation to the air.

rap on the edge of the forge or anvil horn to knock off the dirt, placed in position, and hammered rapidly till all parts are stuck. After the first blow which joins the pieces the thin parts should next be struck to complete the weld, for they lose their heat more rapidly than the thick. If the pieces do not stick at the first or second blow do not continue hammering, as it will only get them out of shape. When two pieces are at a proper welding heat they will stick when touched together.

CASES OF WELDING

Lap Weld. — The easiest weld to make is that where the two ends of a single piece are to be welded together to form a chain link, ring, or the like. This is a lap weld, so called since the ends are lapped over each other.

Making a Chain Link. — The first step is to bend the iron to a U-shaped piece, care being taken to make the legs of even length. The piece is held in the tongs at the bend of the U. The two ends are heated

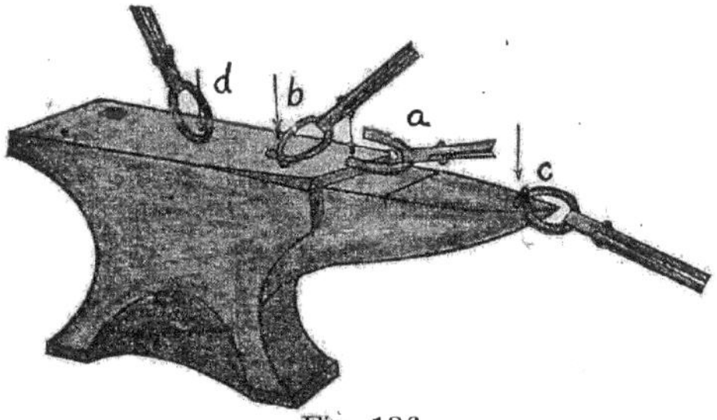

Fig. 136

to a good red and scarfed. The scarf is the roughened bevel made where the pieces lap. It is made by placing the stock on the end of the anvil face where it joins the cutting block (*a*) (Fig. 136) and by giving it one blow. The piece is then moved towards the horn a slight distance and struck another blow. This is continued until the corner is reached. Each blow must be enough heavier than the last so that a bevel will be produced. These blows will make a series of little steps and draw the piece down to a point (*a*) (Fig. 137). The piece is now turned over and the operation repeated on the other leg. Then it is placed over the small part of the horn and bent as shown at (*b*) and (*c*), care being taken to keep the space somewhat angular rather than round as at the other end. This angle should be as sharp as possible and yet fit over the horn without spreading. The points of the scarfs should project slightly. The reason for keeping this end angular can readily be seen from Fig. 138. In each case the portion below the line must be exposed to the welding heat. At (*a*) more stock is below the line: therefore more is liable to injury. The lapped portion of the link is now placed in the fire and a welding heat taken. If the stock is Norway iron, no flux need be used, but if of some inferior grade the ends should be dipped in the flux as soon as it is a bright red and then placed in the fire till at the point of fusion or till a few sparks of burning iron appear. The smith, hammer in hand, should shut off the blast, remove the piece to the face of the anvil, strike a sharp blow on each side of the weld as at (*b*) and immediately place it over the horn as at (*c*), so as to weld the end. It is again removed to the face of the anvil and held as shown at (*d*) and hammered as indicated by the arrow so as to stick and finish the inside. With rapid work this can all be done with one heat. A beginner will probably need two or three; as he should reheat just as soon as there is tendency not to weld, or the iron has cooled below a white heat, for additional hammering will only reduce or thin the

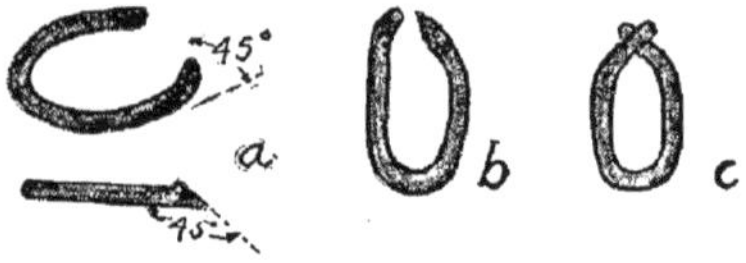

Fig. 137

Fig. 138

stock and make a weak weld. The blows struck on the sides when the piece is first removed from the fire should be as heavy as the piece will stand without flattening the stock.

A second link is made up in the same manner, and these are joined by a third link scarfed and bent as were the first two; only before being closed the two finished links are slipped on as in Fig. 139. The third link is then welded as above described. This will be found much harder, however, as the finished links will insist on getting in the way. To make a chain, sections of three links should be made up and then each section joined by an additional link. The fire must be kept clean and deep and the pieces when heating well covered with coke. The links should be turned often so that both sides will be heated evenly. Another way to make the scarf is to place the link on the face of the anvil after it has been bent to the U-shape and bevel it with the pene of the hammer, as in Fig. 140.[1]

Fig. 139

Fig. 140

Collar. — The collar presents a lap weld with flat stock. The length of the lap should be about one and a half times the thickness of the stock. The scarf should be slightly curved in both directions, as shown in Fig. 141 to make the center of the scarf the highest point. In this way a pocket that would hold slag or air is avoided. In making the scarf the ends are upset

Fig. 141

[1] The author is partial to this method.

a little. The piece is then placed on the anvil and beveled as shown at (*a*) Fig. 142, and finished by rounding with the pene as at (*b*). It must be remembered that the scarfs are to be made on opposite sides of the stock. Fig. 143 shows the ring bent ready for welding. The piece is heated (as in welding a chain link) and all precautions taken to keep the fire clean and deep. Care must be taken to heat the piece slowly so that it will be heated through. The piece should be welded over the horn and special attention given to the thin edge at the ends of the lap, for they cool very rapidly.

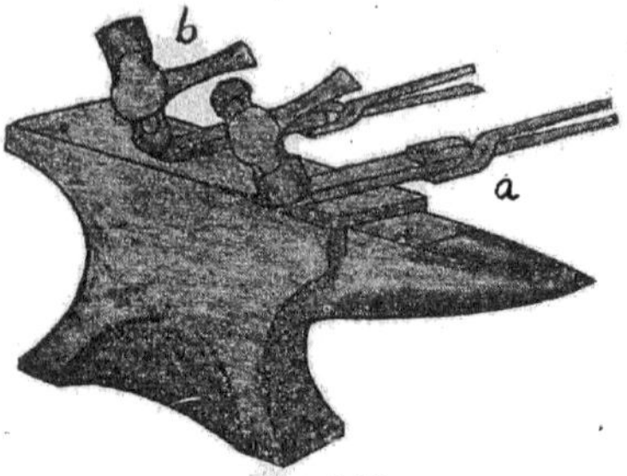

Fig. 142

Washer or Flat Ring. — Fig. 144 shows the stock scarfed and ready to bend. It is upset at the ends and at the same time beveled to an angle of from twenty to thirty degrees. At the first end, the scarf (which is made with the pene of the hammer), is almost full stock thickness on the edge that is to be the inside of the washer and tapers to the outside, the width of the scarf increasing from about $\frac{1}{8}''$ at the inside corner to about 1″ on the outside edge. On the second end, the scarf is on the opposite side of the piece and the full stock thickness is on the side that is to be the outside of the washer. The wide part of the scarf is also on this edge. The scarf should be rounded in the middle to allow

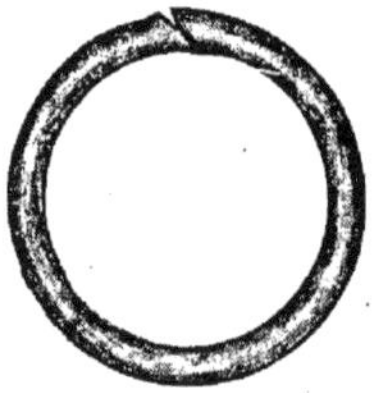
Fig. 143

Fig. 144

the escape of slag, air, etc. A great deal of care must be taken when the stock is bent, to keep it at a red or almost white heat, lest the outside edge crack as in Fig. 145, due to the stretching. The piece is heated for the weld with the same care as described for the other welds. The welding is done on the face of the anvil, the thin edges being taken care of first. After the weld has been made the piece is reheated and placed over the horn to true the outside edge. In all of these welds the blows are regulated to the size of the stock.

Fig. 145

TWO-PIECE WELDING

Bolt Head. — This two-piece weld is about as easy as a single-piece weld, for it is essentially one piece at the time of welding. The stem is upset slightly, (*a*) (Fig. 146). The upset part should be about the same diameter throughout and the distance (*a*) a little longer than the collar that is to be welded on. The collar is made from stock of the proper size or it can be drawn down from larger. In this case assume it is to be for the head of a small bolt and that ½″ round stock is to be used for the collar. Draw down square a portion long enough to reach around the upset part of the stem, and scarf or taper the end as shown at (*b*). Bend this squared portion as in (*c*) to fit tightly around the upset portion of (*a*), and cut off as indicated by the dotted line.

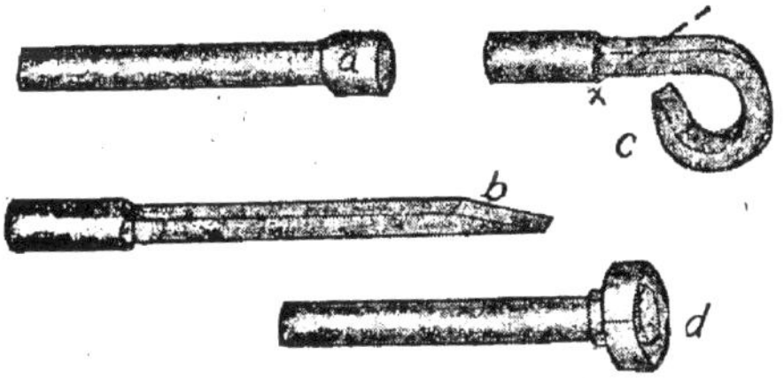

Fig. 146

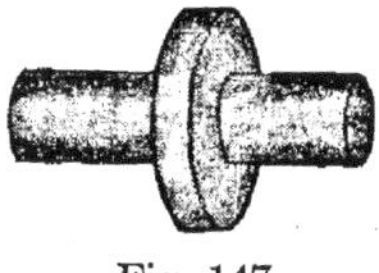

Fig. 147

Close in the end and place the collar on the stem as at (*d*). Bring all up to a welding heat, heating slowly to bring the stem to a proper temperature without burning the ring; place them on the face of the anvil and weld; and finish the rough-looking, irregular top and bottom of the head, by taking another welding heat and dropping the bolt in a heading tool and welding these irregular parts into the head. The head is now in condition to be shaped as desired. Collars (Fig. 147) are welded in a similar manner and finished in a collar swage (Fig. 65). This weld is difficult as it requires considerable skill to get the heat at the proper place, without burning the ends of the shaft.

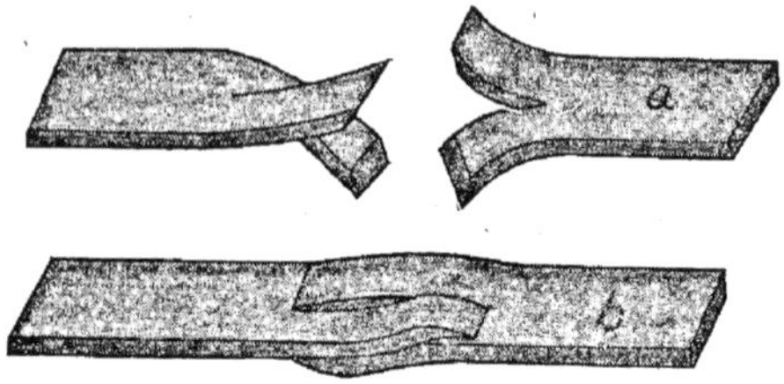

Fig. 148

Split Welds. — Thin pieces that require welding, even though they may be of the proper temperature while in the fire, cool off so rapidly that by the time they can be placed on the anvil they are too cool to weld. This difficulty can be overcome greatly, as follows: split the pieces for about $\frac{1}{2}''$, bend up one part and bend down the other, scarf each split part properly as for a lap weld (*a*) (Fig. 148), place the pieces together and close down the ends as in (*b*). The joined pieces can now be heated as a single piece and placed on the face of the anvil and welded; a flux should be used to keep the ends from burning. This method, is used also in welding spring steel. Fig. 149 shows another type of split weld used on heavy stock.

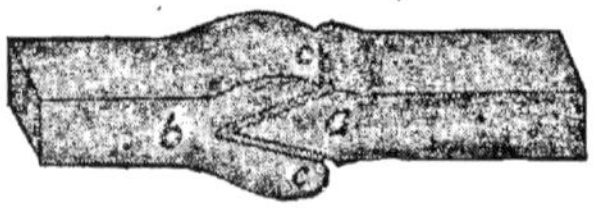

Fig. 149

The ends of each piece are upset. One end is scarfed or pointed as at (*a*) with the side of the bar bulged just back of the joint, while the other end is split and sharpened as at (*b*). The lips (*c*) must be longer than the point of (*a*) so that they will extend well over the bulge. Piece (*a*) is driven into (*b*) and the lips (*c-c*) closed down over the bulge in (*a*) to keep the pieces from slipping. The pieces are now ready to be heated and welded. Care must be taken with the heating to raise the temperature slowly and to use a good flux to prevent the lips (*c-c*) from burning. The piece is placed on the face of the anvil and welded.

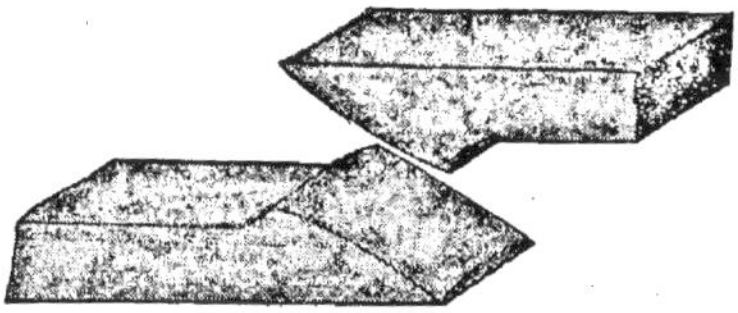
Fig. 150

Lap Weld with Two Pieces. — The ends of the stock are upset and the scarfs are made in a similar way to those for the flat ring. Fig. 150 shows the pieces. The main difficulty with this weld is in handling the pieces. They are placed in the fire scarf side down in as nearly the same position as possible so that both will be brought to the proper temperature at the same time. The hammer is placed in a convenient position on the heel of the anvil. When heated the pieces are grasped to be laid on the anvil with the scarf side of the piece held with the right hand up and the one held by the left down. They are given a rap on the horn of the anvil to free them from dirt and placed as shown in Fig. 151, the

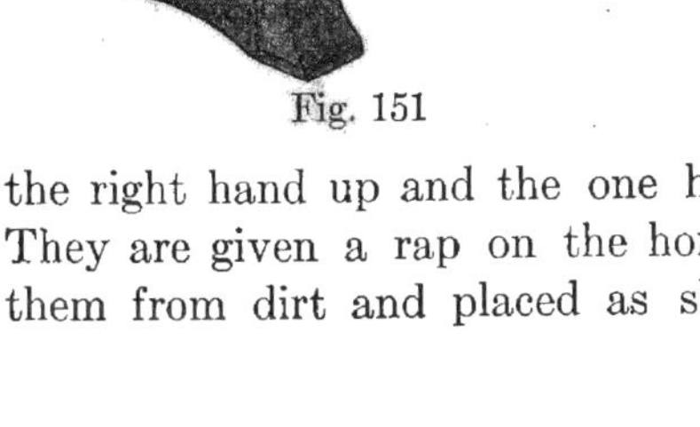
Fig. 151

right hand piece lying on the anvil, the end coming about to the middle of the face, and the left-hand piece resting on the edge of the anvil, the scarf being directly over the scarf of the right-hand piece. The left hand is now raised to press the pieces in contact enough to hold them together when the tongs are removed. The hold on the right-hand tongs is now released so that the tongs will fall to the floor, while the smith takes the hammer and makes the weld by giving a few sharp blows. The piece can then be finished as desired. It may be necessary to reheat. Round pieces can be welded in the same manner. Do not attempt to remove the right-hand tongs in any other way than by letting them fall, for the work might be disarranged or time lost. Don't try to throw them away; simply open up the fingers and let them fall. Have the hammer so that it will be in a natural position for striking when picked up. A beginner should practise taking the pieces from the fire and placing them in proper position on the anvil several times before heating them.

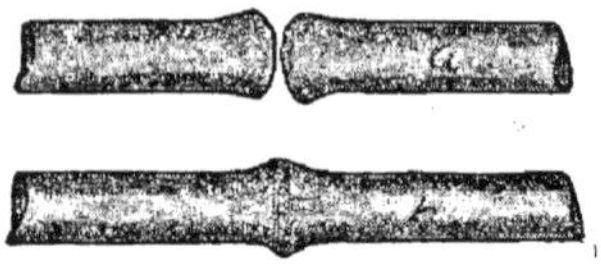

Fig. 152

Butt weld is made by butting together the unscarfed ends of two pieces of stock. In butt-welding the ends should be upset slightly, and rounded so as to force out the slag and dirt (*a*) (Fig. 152). The pieces, when at the proper heat, are united by being forced together. The metal forced out makes a ring (*b*) (Fig. 152), at the joint, which should be hammered back with the hammer and swages to the original size. Short heavy pieces may be

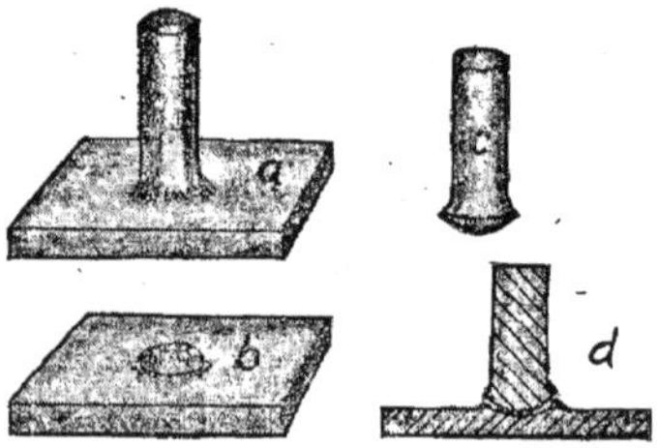

Fig. 153

butt-welded to advantage by a blow of a power hammer. Longer pieces can be jammed together, or, when heavy, the first bar can be rested on the anvil so that the end to be welded will project over the edge a little. The second bar is held in a chain sling, level with the first. The end of the second bar is struck forcibly with a sledge, to drive the bars together to make them stick. Pieces long enough to extend through the fire can be welded in the fire. When the pieces have reached the proper heat, the exposed end of one is held firmly, while that of the other is struck a good blow. Such welding in the fire is aided if the pieces can be supported or guided by a bearing of some sort. As soon as the pieces have been joined they must be taken from the fire and finished to size on the anvil or under the power hammer.

Fig. 154

Jump Weld. — When a stem is to be welded to a head, (*a*) (Fig. 153), the piece is jumped or butted on, making a jump weld. The flat part is scarfed by a small round indentation as shown at (*b*) and the stem upset and rounded as at (*c*). The end of the stem must more than fill the hole but not touch the sides as shown at (*d*). The pieces are brought to the right heat and the stem inserted in the scarf and given a good blow with the hammer. The flange (*e*) is now welded and dressed down with a fuller or set hammer (Fig. 154) or in a heading tool.

Fig. 155

Angle Weld. — Fig. 155 shows the method of scarfing for an angle weld. (*Remember to upset the ends.*) The pieces are heated, taken from the fire, and placed on the anvil in the same way as in the two-piece lap weld, excepting that they are placed at right angles to each other. As the corners are apt to burn during the heating, they should be tipped up in the fire so as to be out of the zone of greatest heat until the main part of the pieces are at a welding temperature. They can be turned down for an instant so as to insure proper heat.

Fig. 156

Tee Weld. — A weld where a stem is to be joined to the center of a piece so as to form a "T" is one of the most difficult of welds, since it is hard to get the proper heat at the center of the top of the "T" without burning the ends. The cross-piece or top should be upset slightly where the weld is to be and scarfed with the pene, (*a*) (Fig. 156). The stem is upset at one end and shaped and scarfed as at (*b*). The bulged portion of (*b*) should be large enough to overlap (*a*) as indicated by the dotted line. To properly heat the scarf of (*a*) without burning the ends can be done best by making a narrow fire which will allow the ends of (*a*) to extend into the banks. The pieces are removed from the fire and welded in a manner similar to that of the angle weld.

Fig. 157

Scarfing Steel. — Fig. 157 shows a method of scarfing pieces of steel to prevent their slipping when being welded.

Welding Steel to Iron. — In welding steel to iron a hollow fire is best. The iron which requires the greater heat is placed into the fire first and heated with borax as a flux nearly to the point of fusion, before the steel face is placed into the fire. The steel is laid alongside of the iron until it reaches a full red or yellow white heat. Since by this time the iron has reached the welding heat, a little more flux is added, the steel is placed in position on the iron while it is still in the fire, and pressed down with the tongs. The iron with the steel now stuck to it is removed from the fire, and placed on the anvil, and the steel is hammered down to complete the weld.

Fagot Weld. — Fagot welding is the welding together of a bundle of short pieces of scrap wrought iron. This is done by placing on a plate or a board the pieces to be welded, arranged in a neat, solid pile. The whole is then securely bound together, placed in a furnace and brought to a welding heat and welded to a solid slab under the power hammer. These slabs can be drawn down to bars of the desired size and shape, and several slabs can be welded together when larger stock is needed.

Welding Steel. — The formation of a soft, pasty surface to steel without burning it is absolutely necessary to effect a union of two pieces of steel. To prevent oxidation, it is absolutely necessary to use a flux such as borax, clay, potash, soda, sand and sal ammoniac, or ordinary red clay dried and powdered, which is a good and cheap flux for use with steel. Borax, when fused, powdered, and mixed with sal ammoniac, is the best known flux, but as it is expensive its use is confined to the finest steel or alloy steel that will not permit of being heated as high as do low-grade steels. Barite or heavy spar makes a very good flux and costs only about half as much as does borax. It does not fuse as easily as borax but forms an excellent covering for the steel to prevent oxidation.

Although steel can be welded, it should be avoided when the pieces are to be hardened, for they are almost sure to crack at the weld when dipped in the hardening solution.

Hammer refining in steel welding is a very important part of the operation and never should be neglected.

QUESTIONS FOR REVIEW

What is welding? What is meant by the welding heat? How do each of the materials, wrought iron, mild steel, and tool steel look when at the welding heat? What kind of metals can be welded? What characteristics do they have when being heated to the melting point? What causes the difficulty in welding? How must the fire be kept? What does air, dirt, and scale do in a weld? How can the scale be prevented from forming? What is a flux? Name some fluxes. Can all fluxes be used on all materials? Can a weld be made without a flux? How does a flux assist in welding? If pieces are not right for welding will hammering make them stick? What will happen if the iron is heated too high? Must the iron be heated fast or slowly? Why? Does it make any difference if the piece is not heated uniformly? When should the flux be added in making a weld? What is a lap weld? How is a chain link made? Why do we make two links separately first and then join them by a third link? Why not each link to the other? How is a flat collar made? Why are the scarfs slightly rounded? How is a washer made? Why must it be kept hot when bending? Why are the ends upset? Tell how to weld a bolt head to a stem. Why are split welds used? What are the different kinds of split welds? Give the operation of taking two pieces from the fire for a two-piece weld. Why should the tongs be allowed to fall instead of being taken from the work? What is a butt weld? What are the different ways of making one? What is a jump weld? Why must the upset part of the stem more than fill the depression in the head? Why must the stem not touch the sides before the weld is made? What must be looked out for in making an angle weld? What makes a T-weld difficult? How can this difficulty be helped?

Name a good way of scarfing steel. What kind of a fire is used in welding steel to iron? Tell how steel is welded to iron. What is fagot welding? What is the important thing to watch in welding steel? Why should steel that is to be hardened not be welded? What is hammer refining?

CHAPTER X

WELDING PROCESSES

ELECTRIC WELDING

There are two principal methods used in making welds by electricity, i.e.: (*a*) The arc system, in which the weld is usually made by fusion; (*b*) the incandescent system, in which the metal is heated only to the plastic condition by electrical resistance.

Arc Welding. — There are three processes of arc welding:

1. The Zerener, in which the arc is drawn between two carbon electrodes and the heat from the arc then directed upon the metal to be welded, which is brought to the melting point and in cooling completes the weld.

2. The Bernardos, in which the arc is drawn between the metal and one carbon electrode, that is, the arc reaches from the metal to be welded to the carbon. The arc may be continued long enough to melt the metal, or it may be stopped at the point of plasticity if a pressure weld is desired.

3. The Slavianoff, in which the arc is drawn between the metal to be welded and one metal electrode. The work is made the positive pole, thus heating the ends to be welded to the melting point. The electricity also causes the metal electrode to melt and flow into the joint, which produces a weld by fusion. The flame of the electric arc produces the highest temperature known, often reaching 7200° F. Some salient points to be remembered when making an arc weld are: the metal to

be welded should be free from dirt or rust. The weld should be well hammered before it cools. A flux is necessary. The work should always be the positive terminal. The arc should be as long as possible. The arc should be given a rotary motion by the hand.

Resistance Welding. — The principle of the incandescent method of welding is that, while the parts to be welded are held tightly together, a heavy current of electricity, at a very low voltage, is passed through the metal and across the joint. The size of the metal at the weld, however, is usually too small for the large amount of electricity to pass through without heating it up, and this occurs so quickly that a welding heat is reached before there is time for much loss by radiation, or damage by oxidation, of the welded parts. It will be noted that this is welding by plasticity, as in the case of the ordinary hand weld.

This resistance system of electric welding differs from all other methods in that the heat is generated in the metal itself. This eliminates the possibility of the weld being defective from the dirt and sulphur of a forge fire, or from the rapid oxidizing effects of some other methods of welding. A flux is therefore usually unnecessary.

There are many ways of fusing the electrical resistance process of welding, of which may be mentioned the following:

Butt Welding. — Here the parts to be welded are arranged end to end and held tightly in place by clamps. These clamps serve to squeeze the joint together as the metal is heated up and made soft. Then the weld is completed by sufficient pressure being exerted on the clamps to upset the joint. This pressure amounts to about 1800 lbs. per square inch for a weld of tool steel. The upset portion of the weld, called the "burr," should be left in place whenever practicable, because of the additional strength it lends to the joint. Much of the electric butt welding is done by welding machines. They work automatically, and require only a few seconds to make a weld. In the case of light work like flat bands, the weld is made in about five seconds, and for 2″ round iron less than 2 minutes is required, while the cost for say $\frac{1}{4}$″ round iron welds is only about 35 cents per 1000 welds, and for 2″ round iron about $65 per 1000 welds.

Lap Welding. — This form of weld is obtained by lapping one piece of metal over the other, thus gaining a stronger weld due to the greater area of contact in the joint. In lap welding by electric machines, the two sheets of metal are overlapped about $\frac{1}{8}$″, and the joint is heated to the point of fusion by a revolving electrode. At the same time the two sheets of metal are squeezed together so that the welded portion is flattened down to about the thickness of the original sheet.

Spot Welding. — It is often necessary in practice to fasten two or more sheets of metal together by drilling holes and then riveting through them. Spot welding does away with this practice. The machines for spot welding are constructed so that the electrodes are in the form of two punch-shaped jaws. The metal to be welded is placed between these jaws, which are forced tightly together. The electric current is turned on with a foot switch, and the joint heated to the welding temperature only over a spot the size of the ends of the jaws, which requires but a few seconds of time. The pressure is released, and the two pieces of metal are welded together at the spot, which cools almost instantly, because the heat was confined to the surface under the electrodes. Then the joint is moved along so that more spots may be welded. A spot weld is frequently stronger than a riveted joint, especially in the case of very thin metal.

Point Welding. — While spot welding is intended to take the place of riveting, point welding is especially applicable to the fastening together of sheets of metal which completely or almost cover each other, and are not held just along an edge as in spot welding. Point welding is done by making a number of raised points on the surface of a sheet of metal and, after placing the adjoining sheet upon these elevations, the current is applied at the points, heating them to a temperature high enough to weld them fast to the upper sheet, which is being pressed upon the points by the electrodes of the electric welder.

Ridge Welding. — This method of making an electric weld differs from point welding in that a long ridge is raised on one sheet of the metal to form the bearing surface for the adjoining sheet, and the shape of the electrodes of the electric welding machines have to be made accordingly. The advantage of this form of weld lies in the strength obtained from the greater welding area supplied by the ridge.

T, L and X Welding. — The tee, ell and cross forms of welds are used in angle-bar work, at intersections, for reinforcing corners, etc. The electric process successfully applies to this class of weld, and the work is quickly accomplished by the electric welding machine. These difficult shapes of welds give a good idea of the wide scope of the electric process of welding.

Chain Welding. — One of the latest applications of electric welding is to be found in making steel chains. In the first attempts the links were shaped in halves, and their ends butt-welded together. This left a troublesome burr in the middle of each link. Then an electric welder was designed which joined the link by a weld on one side, which also left a projection on the link that required to be ground off. Now chain links are successfully welded by automatic electric machines, the weld being made on one side of each link so perfectly that no projection or fin is left on it. The electric welding of chain links has made it possible for heavier loads to be put upon the chains, due to the greater efficiency of the electric over the hand weld.

AUTOGENOUS WELDING

Oxyacetylene Welding.—In the autogenous process of welding the heating medium is usually the oxyacetylene torch in place of the forge fire, although in some instances the oxygen has been used in combination with other combustible gases than acetylene, such as Pintsch gas, hydrogen, etc., but these have not proved as efficient as the acetylene. The method used in making an ordinary autogenous weld is as follows: the two pieces of metal to be joined are laid slightly apart. If the metal is over $\frac{3}{16}''$ thick the ends should be beveled on one side for pieces of ordinary sizes, and in the case of thick heavy metal both sides of the joint ends should be beveled. The flame is then played upon the ends so as to melt the surface and form a little pool of metal. A small rod of adding metal is held in the flame and caused to melt and drop into the pool between the ends. This added metal finally fills up the groove of joint and the weld is completed. It is to be noted that this is welding by fusion, that is, by melting the metal, whereas the ordinary hand weld is made at the point of greatest plasticity of the metal, which is at a temperature about 200° F. lower than the point of fusion. The beveling of the ends of the job is necessary in order to convey the heat to the center of the pieces of metal before the flame burns the outer surface.

The added metal must be of the same kind as the pieces being welded, that is, if the latter are machinery steel the added metal should be machinery steel.

The heat is supplied as follows: the oxygen is confined in a steel cylinder under a pressure of 1800 lbs. per square inch in the high-pressure system, or under a pressure of 300 lbs. per square inch in the low-pressure system, the advantages of the two being that the high-pressure is the most economical system from the standpoint of cost of transportation, and the low-pressure system is most advantageous when the oxygen is generated on the premises of the user. Another tank is filled with acetylene, which is produced directly from calcium carbide in some cases, but when the outfit is to be portable the acetylene is usually obtained by placing asbestos in a cylinder into which is poured acetone. The acetone absorbs the acetylene gas, separating it from the liquid. The acetylene gas then passes through a reducing valve and a hose to the torch, where it combines with the oxygen, forming a flame at the tip of about 6000° F. To make a strong weld the filling in of the joint should be continued until it assumes a convex shape, care being taken to see that the elevated portion makes a thin curved or fillet connection with the main surface, and not a sharp corner which might come from balling the pool too much as it cools.

Autogenous welding is being used extensively in shop practice. It has the unusual advantage of being applicable to the welding of non-ferrous metals such as brass, copper, aluminum, etc., as well as all the ferrous metals like steel, wrought iron, and even cast iron.

Oxyacetylene Building up. — The oxyacetylene flame is being used by many plants in making repairs, by enlarging a worn machine part so that it can be continued in use, thus avoiding the cost of a complete new part. For example, the latest railroad shop practice is to use the building-up process on such work as the repairs of worn piston rods. In this case the worn end of the rod is heating up in a forge fire, and then the oxyacetylene flame used to melt enough steel rod over the surface of the piston rod so that it can be again turned to a fit. In a job of this kind there is a saving of about $10 on each piston rod repaired. This building-up process may also be used to repair a broken bearing, put a boss on a frame, replace a broken gear tooth, etc.

Oxyacetylene Cutting of Metals. — Another very important use of the oxyacetylene flame is that of cutting metals. Indeed this method has become so popular that it has in many instances superseded machines which have long been depended upon for the severance of metals.

In this process the torch tip used has three holes, the center one for oxygen alone, and each of the other two for the combined oxygen and acetylene gases. The cutting is accomplished as follows: A small spot of the metal is made red-hot by the two outside heating flames of the torch.. Then the central jet of oxygen is directed on the hot spot and, because of the great affinity red-hot iron has for oxygen, there occurs instantaneous oxidation of the metal, and a groove is burned through the piece. That is, the metal is not melted, but oxidizes and flies apart in small pieces of scale, like that found near an anvil. The kerf (groove) caused by this disintegration of the metal may be as narrow as $\frac{1}{64}''$ for small work, and a depth of cut of 24 inches has been accomplished. The cutting may be done quite rapidly, the usual speed being about one lineal foot per minute for steel one inch thick.

Steel and wrought iron are the only metals which can be cut by the oxyacetylene flame, which therefore excludes cast iron and the non-ferrous metals.

THERMIT WELDING

The thermit processes were discovered by Dr. Hans Goldschmidt in 1898. Thermit is a mixture of a metallic oxide, sulphide or chloride with finely divided aluminum, and while iron thermit is probably the one most often used, other mixtures may be obtained such as nickel thermit, manganese thermit, chromium thermit, titanium thermit, etc.

The principle of the operation of thermit is that, because of the great affinity for oxygen possessed by aluminum, a reaction is produced upon igniting the mixture in one spot, which instantly reduces the whole mass to liquid. This reaction of the aluminum will entirely extract the metallic element from the oxide, that is, in the case of the ordinary thermit, the iron is separated from the oxygen in the scale and becomes liquid steel.

Thermit is used in many ways, of which may be mentioned: (*a*) The making of welds by fusion, (*b*) the making of welds by plasticity, (*c*) the welding of castings, (*d*) the strengthening of castings.

Thermit Welding by Fusion. — A thermit fusion weld is made in the following manner: assume an engine steel crank-shaft is to be welded. The two parts of the shaft are clamped down to keep them from getting out of line, with their ends about an inch apart. Around this joint is formed a wax pattern. A sand mold, held securely in place by sheet-iron frames, is then shapened over the wax pattern, and as the sand is moist it must be quickly dried out by a gasolene preheating torch introduced into an opening at the bottom of the mold. A crucible is next placed so that the tapping hole in the bottom of the crucible is directly over a pouring gate in the mold. Then the charge of thermit is put into the crucible and ignited. This ignition is accomplished by placing a small quantity of magnesium powder on top of the thermit and lighting it with a match. In about half a minute the whole crucible full of thermit powder will become liquid steel having a temperature of about 5400° F. Then the tapping pin in the bottom of the crucible is struck an upward blow, which releases the molten thermit so that it flows into the mold and fills it up after burning the wax pattern out of the mold. The ends of the shaft are melted by the thermit, and the whole mass fused together. This is literally casting a weld.

By making the wax pattern large enough the welded joint may be so proportioned that it will have greater strength than the shaft itself.

It is best to preheat the parts to be welded before applying the thermit, or the liquid thermit may be chilled enough to prevent successful fusion.

Thermit Welding by Plasticity. — A good example of this method of using thermit may be found in the welding of pipe joints. In this case the ends to be welded are turned up with a hand-facing machine so that the surfaces fit closely together. Screw clamps are fastened on the pipe at each side of the weld. A cast-iron mold is placed around the joint. Then a charge of thermit in a hand crucible is ignited, and as soon as the seething ceases the molten thermit is poured from the top of the crucible into the iron mold and around the pipe ends, which heats up the joint to the plastic condition.

The soft ends of the pipes are then forced together by screwing upon the clamps. This upsets the weld slightly on the outer surface, but the inside diameter of the pipe remains unaffected.

It should be noted that the principles represented in the two types of welds just described are quite different. When welding by fusion the molten thermit is tapped from the *bottom* of the crucible, thus sending into the joint a quantity of liquid steel which melts and becomes part of the steel shaft to be welded. This molten steel serves as a heating medium and then amalgamates with the steel it has melted to form a homogeneous mass.

The weak, molten slag is prevented from getting into the mold by the thermit steel, which, being the heaviest and poured from the bottom of the crucible, is the first to enter and fill up the mold.

When welding by plasticity, however, the liquid thermit is poured from the *top* of the crucible, and the slag, being the lighter and therefore to be found floating on top of the thermit steel, is the first to enter the mold. There it coats all the surfaces of the mold and pipe ends, so that when the thermit steel follows at a temperature of 5400° F. it is kept from contact with the metal surfaces. Its heat is, however, transmitted through the slag jacket to the pipe ends, heating them to an ordinary welding temperature. Were the molten thermit steel to touch the pipes, a hole would instantly be burnt through their surface. The proper amount of heat to be supplied to a weld is, of course, regulated by the quantity of thermit powder placed in the crucible.

The advantages claimed for pipe welding by the thermit process are that the joints are permanent and non-leakable, and cost less to make than buying and installing the ordinary flange connections. The apparatus for these welds is light, and may easily be carried to the job. No outside heat or power is needed.

Thermit Welding of Castings. — To weld a casting has heretofore been rather an unusual operation. It is now done successfully by the use of thermit. In this process the broken joint must be cut out so as to leave a space of about $\frac{1}{2}''$. Then the job is clamped down, a wax pattern made, and a sand mold placed around the joint, practically as has been described for thermit welding by fusion. The parts to be welded must be preheated with a blow torch to a bright red. When the molten thermit is tapped from the bottom of the crucible and allowed to flow into the mold, it melts the wax pattern and fills up the space the wax occupied, fusing to the solid parts of the casting. This leaves the casting stronger than the original section, if the area of the joint can be enlarged by the thermit.

Thermit Strengthening of Castings. — Thermit is also used quite extensively in the repairing of castings having flaws, like blow-holes. If the flaw is small, not over three inches in area, the thermit Pouring-Cup Method may be used to make the necessary repair. First the bad part of the casting should be made red-hot. A pouring cup shaped like an ordinary steel sleeve, made of dry sand, is placed over the flaw, and filled with thermit. The thermit is lighted by the use of a little ignition powder and a match. The reaction takes place in the cup, as it did in the crucible in welding, and the molten thermit fuses the casting, and also fills up the flaw cavity. It is best to pour enough thermit to leave a small elevation beyond the surface of the casting, which may later be ground off.

Another application of thermit to castings is where it is put into the molten iron tapped from the foundry cupola. The thermit powder is packed in a can which in turn is fastened to an iron rod. After the iron has been drawn from the cupola and is in the ladle, the can of thermit called the "little devil" is pushed down into the molten iron. The reaction of the thermit occurs instantly, and as its temperature is about 5400° F. as compared to say 2200° F. for the molten iron, the iron is materially increased in temperature and fluidity, which means more perfect and stronger castings.

WELDING WITH LIQUID FUEL

One of the latest methods used in welding is the oil-fuel furnace. Here an attempt is made to improve on the ordinary soft-coal forge fire for heating up the parts to be welded by the use of an oil furnace. This method is claimed to possess such advantages as: (*a*) A more even heat in the furnace than can be obtained with any other fuel; (*b*) a temperature of heat which is uniform throughout; (*c*) economy of fuel, as the cheapest crude and fuel oils can be utilized.

The oil fuel is supplied to the furnace through a special burner under air pressure, which atomizes the oil so that it readily unites with the air in the furnace and goes into instant combustion. The furnace is made with two chambers, one for the combustion of the fuel, and the other to hold the metal to be welded. This plan prevents the flames from coming in direct contact with the job, and does away with the rapid oxidation so common to the open fire. A flux is therefore unnecessary in oil-furnace welding. This greatly facilitates the work, and in a case on record it was possible to weld the safe ends on locomotive boiler flues at the rate of 60 flues per hour as compared to 16 per hour by the coal fire.

The oil furnace is much used for "blooming" large axles or shafts. That is, piles of scrap iron are placed in the furnace and after being quickly heated to a welding temperature, each pile is taken out and hammered down to a solid piece called a "bloom," from which are made the shafts.

QUESTIONS FOR REVIEW

Describe the Zerener Electric Arc Welding system. How does the Bernardos arc system differ from the Zerener system? What kind of electrodes are used in the Slavianoff system? Is the arc weld made by fusion or plasticity? How does Electric Resistance Welding differ from Arc Welding? How is an electric butt weld made? What is the advantage of the electric lap weld? Where does electric spot welding apply in practise? Describe electric point welding. How does electric ridge welding differ from point welding? Where should electric cross welding be used? How are chains welded by electricity? What is meant by autogenous welding? What is the difference between the high-pressure and low-pressure systems of autogenous welding? What kinds of metal can be welded by the autogenous process? Describe the oxy-acetylene building-up process. What advantage is obtained from its use? How are metals cut with the oxyacetylene flame? Can cast iron be cut with this flame? What is the principle of the cutting? How many jets are used in cutting with oxyacetylene? What is thermit? How is a thermit weld made by fusion? How is a thermit weld made by plasticity? How does the temperature of molten thermit compare with that of the oxyacetylene flame? Describe the thermit method of welding castings. How can a casting be strengthened by the thermit process? Describe a liquid fuel weld. What advantage is obtained from the use of the oil furnace for welding?

CHAPTER XI

BRAZING

Brazing is the joining together of two or more pieces of metal, either similar or dissimilar, by means of a brass spelter.

Hard Soldering is a similar operation, in which an alloy of silver is used.

The Principle of Brazing is that a brass spelter, when placed on metals of higher melting points, as copper, iron, etc., will melt and weld itself to these other metals before they reach their melting point, and if taken from the fire and allowed to set will hold the pieces together firmly. Only enough heat should be applied to cause the spelter to run. Pieces to be brazed must be clean; that is, free from scale, grease, and other substances that will prevent the spelter from adhering to the pieces. A good flux must be used to prevent the surface from oxidizing.

Flux. — Borax is the flux most generally used, for all kinds of brazing. For commercial work granular boracic acid is sometimes used as it is much cheaper. A mixture of borax and boracic acid work well together.

Spelter. — An alloy of equal parts of copper and zinc is the brazing material most commonly used. Often work requires an alloy that is either harder or softer than spelter, which gives rise to the following compositions:

Alloys	Tin	Copper	Zinc	Antimony
Hardest	0	3	1	0
Hard (spelter)	0	1	1	0
Soft	1	4	3	0
Very soft	2	0	0	1

In commercial work spelter mixed with borax, or boracic acid, in proportions found best suited to the work, is found convenient. This mixture should be powdered and mixed thoroughly. In this condition it can be placed on the heated work and it will melt and "run" very easily.

Preparing Pieces. — The pieces must be fitted together with utmost care to get a close joint, since a close joint means a strong braze. It is not necessary to have a place for the spelter, for no matter how tight the joint is made it will find its way between the surfaces.

Cleaning. — The surfaces must be thoroughly cleaned. This is done usually by filing, grinding, or scraping, and the fitting is done at the same time. The joints are always cleaned with emery-cloth or a wire brush. Acid is sometimes used, but if allowed to remain on the work it eats the iron under the braze and causes a weak spot. Acid should never be used without the use of an alkali afterwards to neutralize any active acid which may have remained on the iron. Since alkali will interfere with the braze, unless it is one the bad effects of which can be counteracted by the flux, its use should be avoided if possible.

Methods of Fitting. —Two methods are extensively used in making the joints of a braze (Fig. 158). (*a*) is used for rings of round section and (*b*) for general work. Where the faces are large the butt joint can be used and often surfaces require special fitting or joining.

Fig. 158

Heating. — The source of heat for brazing can be anything that will heat the parts so that they will melt the spelter, as a forge fire, a gas flame, etc.

In the case of the forge, charcoal makes the best fire, and coke is very good. Bituminous coal should not be used until after it has been coked, and the sulphur driven off. The fire should be made into a sort of crater, and allowed to burn hard till all is a good bright red. Then the blast should be reduced. The pieces are placed on the bed of coals if large, or held above it if small, and when the work has reached a red heat, the flux and spelter placed on carefully at the proper spot, with a long-handled spoon having a small bowl. It is best first to add the borax; and when it starts to fuse and run, to add the spelter; later a little more flux when the spelter gets hot. A wire should be used to poke the spelter into the proper place. The work should heat slowly and evenly. As soon as the spelter has run, the pieces should be taken from the fire and allowed to cool. If the pieces are of such a nature that the unused spelter can be rubbed off before the pieces are cool, this should be done, as it is much easier to remove the hot spelter than it is to file off the cold.

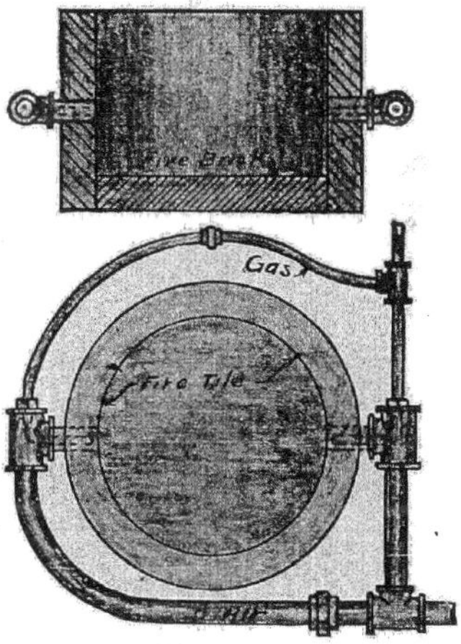

Fig. 159

Brazing furnaces fired by gas can be obtained on the market, or a very serviceable one can be made out of fire-brick or tile. Fig. 159 shows one made from small circular tile. The gas torch is made up of regular stock fittings. The gas flame heats the tiles intensely hot, the heat being reflected on the work, to heat it and melt the spelter and flux. The flux and spelter are added in a similar way to that followed when a forge fire is being used.

The Gasoline Torch. — A very convenient heating device for small work is shown in Fig. 160. The work is placed on, and surrounded by, fire-brick or hard charcoal, and the flame of the torch directed to heat the work by direct contact and reflected heat. Fig. 161 shows the torch in use, brazing a small ring in the hollow of a piece of charcoal. Flat pieces are often clamped between two pieces of charcoal. Fig. 162 shows a band-saw fixed for brazing in this way.

Fig. 160

The Blowpipe (Fig. 163) replaces the torch nicely in either of the cases above.

The small blowpipe, an alcohol lamp, or even a candle, is useful in cases of small work. The work is placed on charcoal, as in Fig. 161, and the flame of the lamp is impinged against the coal effectively to heat the work and melt the spelter.

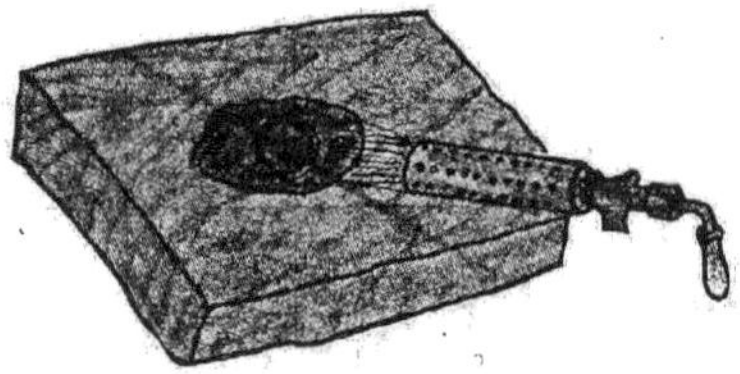

Fig. 161

Hot Tongs. — Extra heavy blacksmith tongs are often heated red-hot and are used to heat the work to be brazed by the gripping of it at the proper place. In this case the work necessarily must be thin to be heated rapidly. The flux and spelter should be placed between the pieces before the hold with the tongs is taken. The mixture of powdered flux and spelter spoken of at the first of this chapter is best for this case.

Brazing by Immersion consists of dipping the work in a bath of melted spelter, on top of which floats melted flux (Fig. 164). The melting can be done in any crucible suitable for the size and shape of the work. The piece is first dipped into the melted flux and held there till it is heated sufficiently. Then it is passed into the spelter which flows into the openings to be brazed. As the molten spelter will adhere to any other parts of the metal it may touch, it is necessary to coat the work with a graphite paste wherever the spelter is not wanted. This can be done easily and quickly, thus there will be little or no spelter to clean off. This makes a saving in time in cleaning as well as in the brazing operation itself. It seems that any of the following fluxes work well with this method: borax; 1 part borax and 3 parts boracic acid; or 3 parts borax and 1 part boracic acid; boracic acid alone, and soda mixed with borax.

Fig. 162

Fig. 163

Cast Iron can be brazed by the aid of several patented preparations which are on the market, or by the Pitch method. In this method the surfaces to be brazed are cleaned carefully and covered with a coating of oxide of copper mixed with a liquid, such as sulphuric acid. These preparations will allow the oxide to be spread on and hold it on after it dries. The oxide reduces the carbon in the cast iron, and allows the spelter, which is applied as in brazing other metals, to take hold and join the pieces.

Fig. 164

Without this oxide of copper, the carbon in the iron would act as does the graphite coating in the dipping process, preventing the spelter from sticking to the iron and from joining the parts.

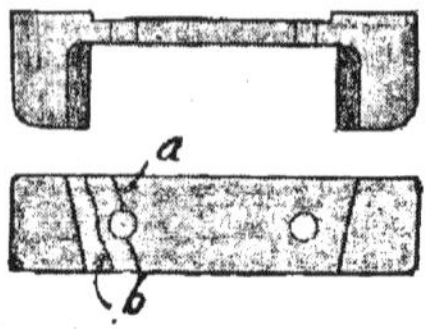

Fig. 165

As the tensile strength of the spelter is greater than that of the cast iron, the brazed piece will be stronger than it was before the break. This is well illustrated in the piece shown in Fig. 165. This piece was brazed by the author and when put into service was again broken; but instead of breaking through the brazed part (*a*), which was the weakest part of the casting, it broke through a thicker part at (*b*).

QUESTIONS FOR REVIEW

What is brazing? What is hard soldering? What is the principle of brazing? What fluxes are generally used? What is spelter? How should pieces be fitted for brazing? Is it necessary to leave a space for spelter? How are pieces cleaned? How many methods of fitting? How may pieces be heated? Describe a brazing furnace. How is a gasoline torch used? A blowpipe? On what kind of pieces can hot tongs be used? Describe the operation of brazing by immersion. What does the graphite coating do? How may cast iron be brazed? What is copper oxide used for? Does brazing make a strong joint?

CHAPTER XII

CARBON TOOL STEEL

The forging of tool steel differs but little from that of wrought iron, except in the extreme care with which it must be heated and handled. Therefore directions for specific operations will not be given, as similar ones already have been described. In this chapter this difference in heating and handling, the action in the cooling bath, and the characteristics of the steel itself will be treated.

Steel. — There are many makes of steel and several grades of each make. The average man, however, recognizes but two kinds; i.e., tool steel and machinery steel. Since machinery steel, or low-carbon steel, closely resembles wrought iron, the method of working it is practically the same. Therefore what has already been stated with regard to wrought iron will apply equally well to machinery steel. Tool steel, however, when cooled more or less rapidly from a red heat or one just above, acts differently. Owing to the sudden cooling the carbon and the iron of the steel form a peculiar chemical compound which makes it very hard. This action will form the subject matter of this chapter.

Tool Steel[1] is steel from which tools requiring a hard cutting edge are made.[2] It usually contains from .5 to

[1] When the term *tool steel* is used, it usually is understood to mean carbon tool steel, unless otherwise indicated.

[2] In the past few years a great many tools have been made from other steels that possess hardening qualities. These steels will be taken up in a later chapter.

1.5 per cent of carbon. This carbon gives the steel the property of hardening, when heated and then suddenly cooled. Tool steel is used for cutting tools or for some parts of machinery where great hardness is required. On the other hand machinery steel is used for parts that do not require hardening at all, or at the most only surface hardening. As it is of a lower grade than tool steel, and is soft, it can be machined or forged easily, heated to a higher temperature, and is cheaper. In fact, in some cases, about the only difference between machinery steel and wrought iron is the process by which it is made.

Temper is the term used by steel makers to indicate the percentage of carbon or hardening elements in the steel. Steels are designated as of low, medium, or high temper, or by some letter or mark meaning that the steel has a low, medium, or high percentage of carbon. Steels are also classed as follows:

Razor Temper ($1\frac{1}{2}$ % carbon). This steel is so high in carbon that it can be handled only with the greatest care, but it gives a very hard cutting edge.

Saw File Temper ($1\frac{3}{8}$ % carbon). This steel will work somewhat easier than Razor temper, but must be kept below a red heat.

Tool Temper ($1\frac{1}{4}$ % carbon). Can be worked at a heat up to the point at which scale begins to form. This steel is used for drills and lathe and planer tools.

Spindle Temper ($1\frac{1}{8}$ % carbon). This is worked at about the same heat as Tool temper, and is used for milling cutters, thread dies, and large planer tools.

Chisel Temper (1 % carbon). Can be worked at a good red heat. It is used for chisels and tools that are required to be tough to stand blows from a hammer.

Set Temper ($\frac{7}{8}$ % carbon). It is worked at the same heat as chisel temper, and is used for press dies or where a hard outside surface is needed with a backing of tough metal to stand great pressure.

Point.—In steel the term "point" means the one one-hundredth of 1 per cent of any element (usually carbon). Thus 100 points means 1 per cent or 60 points carbon steel is steel with 0.60% of carbon. From this arises another notation as follows:

Very hard	150-point carbon.
Hard	120–100 point carbon.
Medium	80–70 point carbon.

Heating the steel is perhaps the most important part of the hardener's duties, hence facilities for accomplishing it should be the best possible for the kind and quantity of work to be handled. If the amount of work is small and can be heated in a blacksmith's forge, the forge will answer. But if the quantity is large or requires special apparatus, special furnaces should be procured, because with them the cost of the work can be made much cheaper, and time will be saved.

If the forge is used either in forging or hardening steel, the fires should be clean and deep enough to keep the blast of air from striking the work. Also the work should be covered with a layer of coal to prevent its contact with the air. Otherwise the oxygen will decarbonize the steel and thus keep it from hardening. Steel will crack from sudden contraction if the fire is so shallow or the steel so placed in the fire that a cold blast strikes it. Especially is this so if the piece has thin projections, which owing to their small size are very susceptible to changes in temperature. If a big piece of steel is to be heated it is necessary to have the fire large enough to heat the piece uniformly.

Charcoal is considered an ideal fuel for heating steel as it is practically pure carbon, but if it is used the fire should be kept well supplied with new coal, or it will be necessary to use a strong blast, which is likely to reach the steel and cause it to crack. It is stated by E. R. Markham in his most excellent work "The American Steel Worker," that high carbon steel will not become so hard on the surface if heated in charcoal fire as if heated in one burning coke. The best way is to heat in such a manner that the steel will not come in contact with the fuel: as in a muffle furnace, a piece of pipe or an iron box. When, however, the work is to be turned in a lathe afterwards, the open fire is better because it heats rapidly.

Furnaces. — The muffle furnace (Fig. 166) is neat and easily managed. It is made to use illuminating gas, and can be procured in almost any size. The steel is placed in the muffle and the gas burned in a chamber which surrounds this muffle, so that the steel is heated by radiation from the walls of the muffle without direct contact with the products of combustion, and, since the door is closed, the steel is protected from oxidation, which would affect the composition of the steel. The gas can be so regulated that a very even heat can be maintained. The furnace should be made so that the work can be seen without opening the door. An ingenious smith can make a very good furnace adapted to burn any kind of fuel.

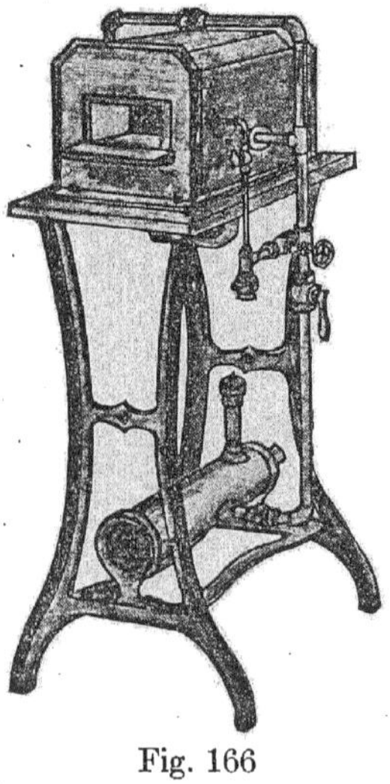

Fig. 166

The Gas Torch (Fig. 163) answers for heating an occasional small piece. The flame impinges upon the fire-brick and reflects the heat onto the piece which is held in the flame. This will heat small pieces very rapidly. Fire-brick can be made into an oven or muffle, and by the use of a torch on each side a very effective heater for small work can be made.

Bunsen Burner is sometimes used to heat small drills.

Heating Baths. — Lead, tin, glass, cyanide of potassium, and other materials are heated in crucibles to special temperatures, and the articles dipped into them. They exclude the air, prevent oxidation and decarbonization of the steel, and produce a very uniform heat. The crucibles can be heated in an ordinary forge, but much more uniform heat can be obtained in special crucible furnaces heated by gas. There should be some means of carrying away the gas and fumes, especially from the lead and cyanide baths. Cyanide of potassium is a very dangerous poison, and should be used with the greatest care.

Lead Bath. — Melted lead, maintained at a uniform temperature, is a heating bath very commonly used. It is melted in an iron ladle, or better in a graphite crucible. The lead should be pure and free from sulphur, as a small amount of sulphur in the bath ruins steel by eating away the surface, giving it an open, sponge-like appearance. The sulphur also makes the steel "hot short." The articles to be hardened are placed in the lead and held there until heated to the proper temperature. They are then removed and plunged into the cooling bath. There is some objection to lead because it sticks to the work, but this can be prevented somewhat if the lead is kept clean and free from oxide or if the work is first dipped into a solution of cyanide of potassium — one pound to the gallon of water — or into a solution of brine. The work must be dried before being dipped into the lead, for the moisture would cause the lead to sputter, and this likely would burn the hardener. Oxidation is prevented by placing a layer of charcoal on the surface of the lead. One difficulty with lead is that steel must be held down or it will float in the bath. The lead bath should be heated to and held at about the temperature at which the steel is to be heated. If it should rise above this temperature it should be cooled by placing a large piece of iron into it.

Cyanide Bath. — Ferrocyanide of potassium heated red-hot in a cast-iron crucible makes an excellent bath for hardening, as it does not stick to the work, has no tendency to oxidation, and the cyanide has a hardening influence on the steel. As with lead it must be chemically pure. Cyanide of potassium is a deadly poison; it should never be used by the inexperienced or melted in any way that would allow the fumes to escape into the room. Fig. 167 shows a furnace that is suitable for melting cyanide.

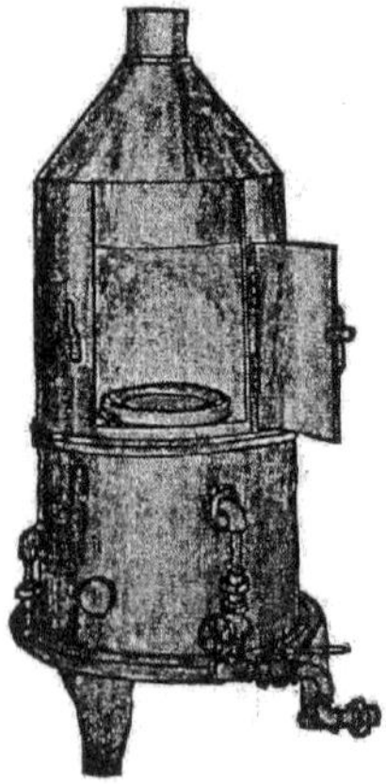

Fig. 167

Uniform Heating. — It should be remembered that pieces must be heated uniformly and that as the thin portions of articles with unequal sections heat more rapidly even in a lead or cyanide bath, the thick portions in some way should be preheated before plunging the whole piece into the bath. This can often be done by holding the heavier portions in the bath until partly heated before the whole piece is plunged.

Location of the Furnace. — The location of the furnace or forge used for heating steel to be hardened is a very important matter, possibly the most important concerning the equipment. It should be placed where direct sun-light or any strong light will not strike either the forge or the operator. The ideal location is a room where the light will be subdued and alike from day to day, which is an important factor in producing uniform results in hardening, for in judging the temperature of the steel by color, the same temperature will not show the same color except under uniform conditions of light. When baths are used it is just as important to look after the ventilation as after the light. The room should be free from dampness and drafts and well supplied with fresh air.

Heating Tool Steel. — The grain of tool steel changes noticeably with slight changes of temperature. When steel is being hardened the lowest heat possible to give the proper grain should be used. This temperature varies with the carbon content, and other hardening elements, as well as with the make of the steel. To get results uniformly hard a lower carbon steel must be heated to a higher temperature than a higher carbon steel.

There is a proper heat for each quality of steel; it generally is spoken of as cherry red. But what is this cherry red? Do any two people see a color exactly alike? The proper color at which to harden a piece of steel is that color which will produce the finest possible grain in the steel — a grain (when cold) that looks like very fine gray silk. At this temperature steel will be hardest. There is a point — the decalescent[1] point —

[1] Decalescence, a phenomenon exhibited by steel when heated to a temperature of about 1400° F., at which point the carbon changes from *pearlite* to *cementite*, or *hardening carbon*. This condition is indicated by the heat curve failing momentarily to show a change of temperature, even though the supply of heat has not been reduced. When the piece of steel is cooling down a similar phenomenon occurs at a point about 50° to 100° F. below the decalescent point where, because of the retransformation of the cementite carbon back into pearlite carbon, the cooling momentarily ceases and the piece of steel actually reglows. This is called the *recalescent point.*

at which the steel for the instant continues to absorb heat but does not rise in temperature. At this point there is a rearrangement of the carbon of the steel. The grain is the finest when the steel is immersed at the decalescent point. If the steel is heated above this temperature the grains increase in size. The larger the grains become when the heat is reduced below the decalescent point the softer will be the steel.

The writer when using a new steel or when teaching the hardening and tempering of steel always tries the following experiment: draw the steel down fairly thin so that it will break easily; heat to some temperature near the one supposedly proper and fix the color well in mind; cool the steel quickly by plunging it into water, break off a small piece, examine the grain and lay the piece aside for reference. Repeat this experiment at several different temperatures, always fixing the color in mind. The color which produces the finest grain after immersion is the color to which the steel should be heated when hardening tools from that bar. When determining the proper color hold the piece of steel in a dark place, as under the forge, in a dark corner of the coal box, or in an old nail keg.

Drawing Temper. — While hardening produces the finest grain, it causes steel to be too brittle for many purposes, and in order to toughen the tools it is necessary to withdraw some of the hardness. This is done by reheating the tools to about 500° F. and then immersing them quickly. This reheating is called *drawing the temper*, or *tempering*.

Rules for Heating. — 1. The steel should be heated uniformly or strains will be set up in the piece, causing cracking when hardening.

2. Do not heat steel faster than it can be heated uniformly. If the heating is forced, the light parts will be

heated much faster than the heavy parts. The fire must be so regulated that all will heat evenly.

3. Do not let a piece heat more slowly than is necessary or soak in the fire, for it will be oxidized, decarbonized, and it will absorb impurities from the fuel.

4. Heat the piece just as fast as it will take the heat.

5. Do not heat the piece above the proper point and allow to cool in the air until the correct color is reached, for then only the outside of the bar is at the correct heat, and the inside will still be hot and will have a coarse grain. It should be cooled to the black and reheated.

6. Steel should be hardened on a rising heat, never on a falling.

7. Turn the pieces over and move them around while heating so that all parts will be equally exposed to the hottest part of the fire.

8. Use special care with round pieces as they seem to crack more easily than pieces of any other shape.

9. If pieces have heavy and light portions, heat the heavy parts first, as the lighter parts will heat partly by conduction. If it is not possible to do this, then heat in a slow fire.

10. Never use a fire in which the blast can strike the steel, for it will crack almost invariably. To prevent this, build a deep fire and keep the steel covered.

11. Do not heat with the steel exposed to the air.

Reheating. — When steel is cooled the outer surface has set before the interior has cooled. This causes internal strains often powerful enough to break the piece with a loud report. To remove these strains it is necessary to reheat the pieces. This can be done by placing them in boiling water and keeping them there until heated through. This has the effect of making the pieces pliable enough to adjust themselves internally.

Annealing steel is softening it so that it can be worked by ordinary cutting tools. Annealing also counteracts strains set up in the piece by machining which would cause it to crack in hardening. The longer it takes a piece to cool the softer it will be. There are three ways of annealing tool steel. One method is to heat the piece to the recalescence point or a little above, allow it to cool in the air till black (in the dark) and then dip it into water, preferably soapy or oily. This is called *water annealing*. The piece should not be chilled by draft or by being placed on cold metal or stone. A second method, called *cover annealing*, is to pack the piece to be annealed in a box of lime, ashes, or fire-clay, which previously has been heated, and allow it to stand until perfectly cold. Lime is the best material to use, but it must always be heated to drive off all dampness so that it will not chill the steel.

Markham gives the following excellent method. Place the heated steel between blocks of wood, pack the blocks and work in a box of previously heated lime (Fig. 168). The pieces of board will smolder and keep the steel hot for a long time.

Fig. 168

The third method, called *pack annealing*, is a very satisfactory method by which to obtain soft steel. This consists of packing the steel to be annealed into an iron box with charcoal or burnt bone. There should be a layer of the charcoal on the bottom, then the steel is placed on the charcoal. The pieces of steel must be at least $\frac{1}{2}''$ apart; the spaces are filled with more charcoal and other layers added till the box is filled. Over the last steel there must be a layer of charcoal as thick as that on the bottom. The cover of the box is then placed on and luted with clay. The box with contents is placed into a furnace, heated slowly till its contents have become red, and then the furnace and its contents allowed to cool slowly. The cover of the box should have a few holes into which iron wires are placed reaching to the center of the box. These wires are drawn occasionally, and when one is drawn that is properly heated its full length, the fire is allowed to burn a few minutes longer and then is extinguished.

Don'ts for Annealing.—1. Don't overheat.

2. Don't subject the work to the heat longer than necessary after it has become uniformly and properly heated.

3. Don't let the temperature become anything but uniform.

4. Don't let the piece heat unevenly.

5. Don't use a packing material that will take the hardening elements out of the steel or introduce impurities into it.

Graphic Representation of the Changes in Carbon Steel.—The diagrams (Figs. 169–170)[1] show graphically the changes produced in the size of the grains of steel and in the nature of the carbon as the steel is heated to the melting and refining points, and slowly and rapidly cooled from these points.

In these diagrams time is measured horizontally and temperature vertically. The circles represent the grains of the steel; the full lines, the hardening carbon; and the dotted lines, the pearlite carbon.

In (*A*) (Fig. 169), we start with the original bar and heat it to the melting point. It will be seen by the circles that the grains remain the same size up to "W," which is the decalescence point. At this point the grain suddenly becomes very small, but as the heat is increased, the grain grows in size rapidly until at

[1] Taken by permission from "Notes on Iron, Steel and Alloys," by Forrest R. Jones.

the melting point the grain is much larger than in the original bar. Observing the lines representing the carbon we see that the carbon remains in the pearlite form up to the point "W," but there it suddenly changes to the hardening form and remains thus to the melting point "*M*." If the piece is suddenly cooled now from *M*, as shown in (*B*), the grain will remain when quenched the large size that it was at *M*, and

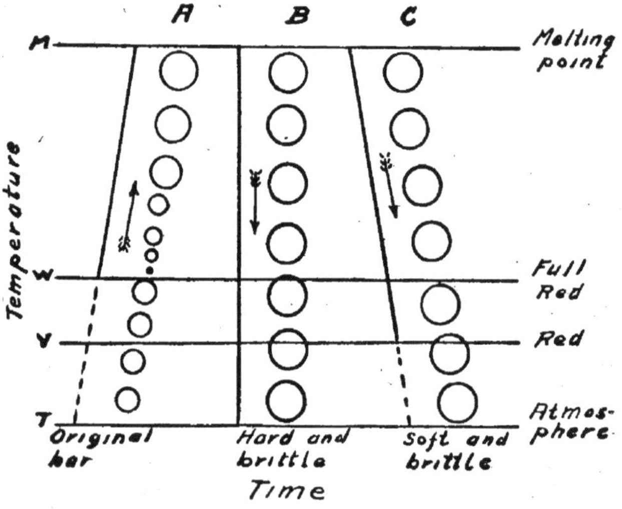

Fig. 169

the carbon remains in the hardening form down to the atmospheric temperature *T*. The steel in this condition is hard, brittle, and worthless. If, however, instead of cooling the piece rapidly as in (*B*), we had cooled it slowly as in (*C*), we would have had the same large grains as at *M* and the carbon would have been in the hardening form until red temperature *V* had been reached, when it would suddenly have been converted to the pearlite form and the piece would have been soft and brittle. It will be seen from this diagram that whenever the steel is heated to the point *M*, no matter how the piece is cooled, the grain will be coarse, and the steel will be brittle and useless.

At (*A*) (Fig. 170), we have the bar heated up to *W*, or a few degrees above the point at which the fine grain is produced. (*B*) shows the sudden cooling from this point to *T*, and we find the grain very fine and the carbon all hardening. In this condition the steel is hard and brittle, but owing to the fine grain it is in the best

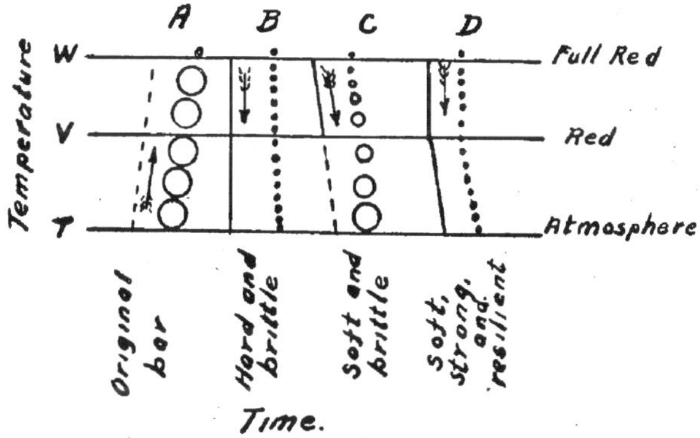

Fig. 170

structural condition for tools. By being slightly reheated when in this condition (drawing the temper or tempering) the grain can be kept nearly the same size and the steel can be made tough. (*C*) shows the effect of slow cooling from *W* to *T*, or annealing. It will be noticed that the grains increase in size to that of the original bar, while the carbon is in the hardening form to *V*, and in the pearlite form to *T*. The steel is soft but brittle, but by being reheated it can be made to take any form possible in the original bar. (*D*) shows the effect of another type of annealing. Here the piece is cooled suddenly to *V* and then is allowed to cool slowly to *T*. The grain will retain the fine structure and the carbon will remain in the hardening form, but the steel will be soft, strong, and resilient.

Hardening Baths. — After steel has been heated to the proper temperature, it must be dipped into a cooling solution. The more rapidly this solution will conduct the heat away from the steel or the more closely it will adhere to the steel, the harder (everything else being equal) will become the steel. Hence cold water will produce a harder steel than warm, and mercury harder than cold water; while substances like oil will leave the steel softer than the warm water. This fact allows us to use various cooling baths to obtain different degrees of hardness and toughness. For general use clear, cool water answers best; it is effective and cheap. It should be clean and dirt should not be allowed to accumulate on the bottom of the tank. It must be remembered, however, that in most cases too rapid cooling must be avoided, for if the outside becomes rigid before the interior is set the piece is likely to be placed under such internal strains that it will crack. For this reason the water is best about 60° F.

Brine. — Another bath very much used is a saturated solution of salt water or brine — made by dissolving all the salt the water will hold.

Oil, such as linseed, neatsfoot, or most any fish oil will produce excellent results on thin tools requiring a hard edge and tough center. They also are used for springs. Tallow, sperm, and lard oils produce great toughness and are used for springs.

Solutions. — The following solution is much used on small cutting tools as taps and reamers:

citric acid	1 lb.
water	1 gal.

Markham gives the following, as being recommended to produce hard, tough tools:

salt	½ teacupful.
saltpeter	½ ounce.
pulverized alum	1 teaspoonful.
soft water	1 gal.

Acids, such as sulphuric, are sometimes used, but they are not recommended, as they attack the surface of the steel.

Water with a Layer of Oil on its Surface is often used to harden high-carbon steel when tools having teeth that meet the body of the piece with sharp angles are to be made. The oil adheres to the steel at the bottoms of the teeth and prevents a sudden change of temperature with the resultant cracking.

Flowing Water. — Some pieces require jets of flowing water to be projected against a certain face, that the face may be hard and the body tough. Die faces are hardened in this way (Fig. 171).

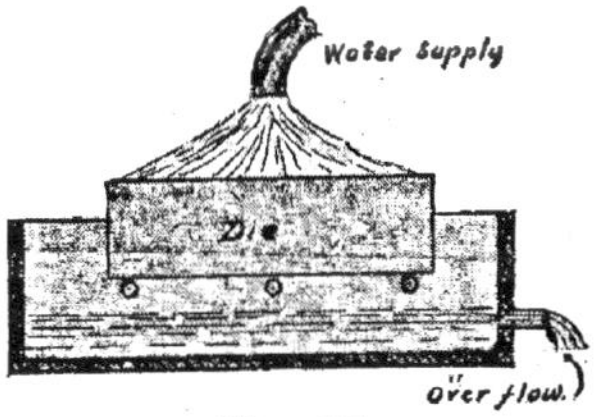

Fig. 171

Tempering Colors. — When polished steel is heated an oxide is formed on the brightened surface which has a color characteristic for each temperature between certain limits. When steel has been hardened these oxide colors indicate a definite temperature and degree of toughness. Hence by heating the steel to the temperature required to give it the degree of toughness needed, or to the oxide color that represents this temperature, and then quenching the steel, we may be certain that it is tempered as required. The temper colors are as follows:

Light straw	430° F.
Dark "	470° F.
Brown	490° F.
Brown with purple spots	510° F.
Purple	530° F.
Light blue	550° F.
Dark "	600° F.

Methods of Hardening. — Articles to be hardened generally can be divided into two classes: *Class one*, those only dipped partly into the cooling bath, the heat in the unplunged portion flowing to the hardened or cooled portion and thereby reheating and drawing the temper; *Class two*, those to be dipped entirely into the cooling bath and the temper afterwards drawn by some method of reheating. The details of the methods used in each class are almost as varied as the work to be hardened.

A few examples will be given which will indicate how to proceed in simple cases.

CLASS ONE

Chisels, punches, most lathe and planer tools, and articles requiring a hard cutting point or edge and a tough shock-resisting body, generally fall under class 1. The cold chisel will furnish a good example of this class and will be studied in detail. By any of the heating methods, generally in the forges, the piece is heated to the proper temperature to a point indicated by the line (*a*) (Fig. 172), about 3″ from the end and is dipped to about 2″ from the end as indicated by the line (*b*) into a cooling bath (water) till cold. The chisel should be moved slightly up and down in the bath to counteract the tendency to crack at the water line, and also back and forth or around in a circle so that the steam formed will have

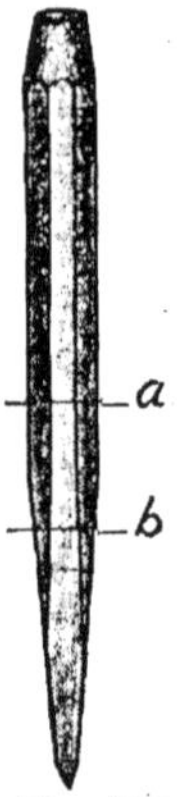

Fig. 172

a chance to escape and allow the bath to come into direct contact with the tool. When the part in the bath is cool, the chisel is removed and the cooled portion polished with an emery stick[1] to remove the black oxide (Fig. 173). This will allow the tempering colors to be seen which indicate when the temper is drawn to the desired point. This drawing of the temper is accomplished by allowing the heat in the portion of the chisel back of the water line (*b*) (Fig. 172) to flow to the cooled portion, there to heat and soften it. This heating will cause the tempering colors to appear; First, the light straw near the water line will appear, and as this moves toward the point the others will appear in order, — dark straw, brown, brown with purple spots, purple, light blue, and dark blue. When the purple or the light blue has reached the cutting edge of the chisel, all further drawing is stopped by plunging the chisel in the bath until it is cool. Should the chisel be cooled so much at the first dipping that there is not heat enough left to draw the temper the required amount, it can be reheated by being held over the fire. Care should be exercised not to have the tempering heat too high or the colors will run too fast and be too close together. In this case the chisel should be dipped again and withdrawn quickly, to take away

Fig. 173

[1] An emery stick is made by gluing emery-cloth to a stick or by covering the stick with glue and sprinkling with emery.

some of the heat and cause the balance to flow more slowly.

Diamond Point (Lathe Tool).—Fig. 174 shows how a

Fig. 174

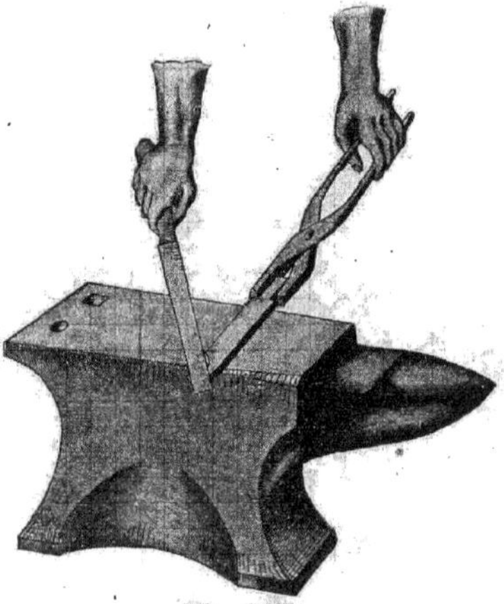

Fig. 175

diamond point tool should be dipped; and Fig. 175, how it should be polished. Most of the other lathe tools should be dipped in a similar manner, and polished the same as the cold chisel on one of the faces extending back from the cutting edge.

Side or Facing Tool. — Fig. 176 shows a method of dipping a side tool for a lathe, which is about the ex-

Fig. 176

Fig. 177

treme of class 1. As can be seen from the figure, only a very small portion of the tool with which to draw the temper is left undipped. This tool is often dipped completely and then the temper drawn by placing it on a heated block of iron (Fig. 177).

CLASS TWO

Taps, Drills, Reamers and Similar Tools should be heated to the refining temperature, in a muffle furnace or in an iron pipe placed in the forge fire (Fig. 178),

Fig. 178

and then plunged into the cooling bath and completely submerged. They must be moved around until cold to insure close contact with the bath. These tools must be dipped as nearly vertically as possible to prevent warping. Before the temper is drawn the scale in the grooves of the taps must be removed, by means of an emery stick or other means, and the surface of the drills and reamers polished so that the colors can be seen. The temper is then drawn by the tools being held in a red-hot ring (Fig. 179) or in the gas-pipe again, without allowing it to touch, or by

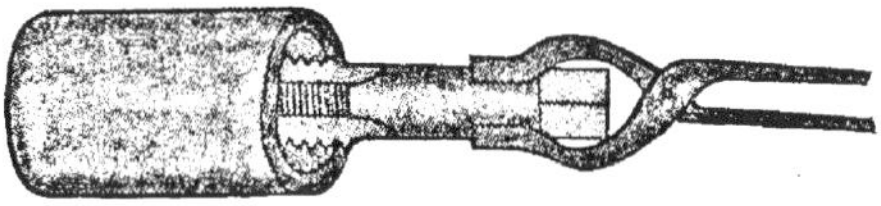

Fig. 179

its being moved around in heated sand till drawn to the proper color, which is a straw.

Shank Milling Cutters if required to be hard all over are treated just as taps, but if the shank is to be left soft, they should be dipped to the dotted line (Fig. 180) and held in the bath at this point until the shank is cool. The temper of the cutting part is then drawn as described under Taps.

Fig. 180

End Mills are treated like the shank milling cutters except when they have deep recesses in the ends such case they should be dipped with the holes up (Fig. 181) and the temper drawn as in taps. If the shank is required to be soft, some special device can be used, as illustrated in Fig. 182. Here a sleeve that fits the shank loosely when cold is heated, slipped over it and held there till the shank is drawn the required amount. The cutter part should be held in water to prevent the heat running down and drawing the temper in it; care should be taken also not to get the heated sleeve too far or too tight on to the shank, as it will shrink in cooling and grip the tool so that it cannot be removed.

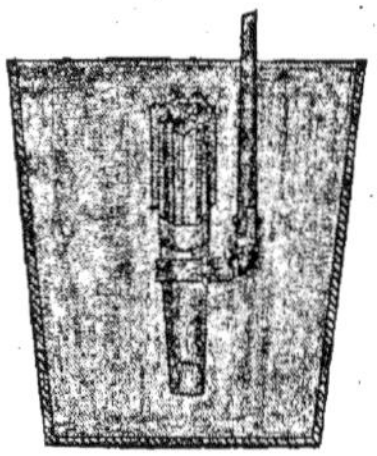

Fig. 181

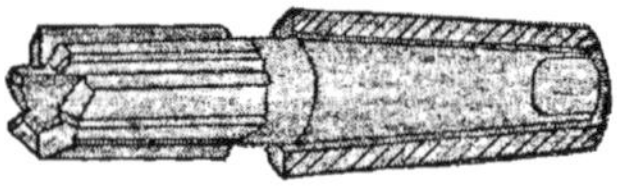

Fig. 182

T'slotters should have an iron wire (*a*) (Fig. 183), as this prevents cracking where the head joins the neck, since it delays the cooling at this point. The cooling should be done by dipping the piece all over and drawing the temper by reheating the shank as was the end mill, or it can be hardened as under class 1. In any event the heat should run from the shank to leave the neck blue and the cutting teeth a straw.

Half Round Reamers are treated in the same way as other reamers except that instead of being dipped vertically they should be held at an angle (Fig. 184), the curved side down, to prevent warping. The shape of the piece will determine the angle at which the piece must be dipped. Knowledge of this angle must be gained by experience.

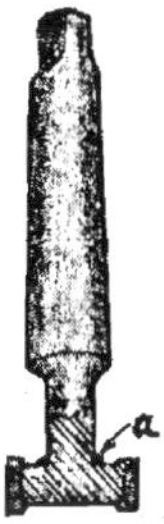

Fig. 183

Milling Cutters should be heated in a muffle furnace or iron tube which has a flat bottom, but if neither is to be had a hollow fire will answer. They should be dipped endwise and completely submerged. If the cutter is large, it should be taken from the water before it is entirely cool and the cool-

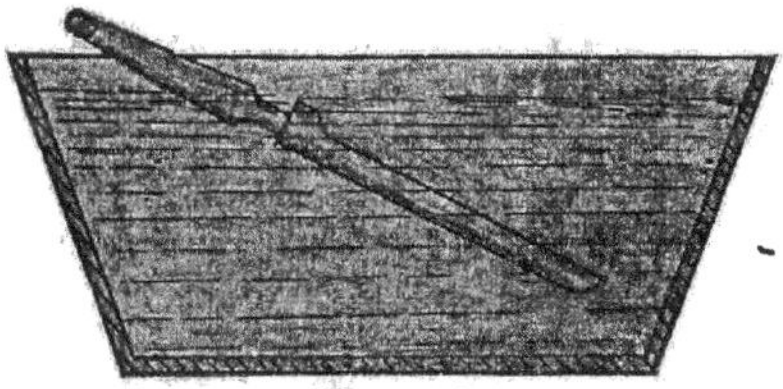
Fig. 184

Fig. 185

ing should be finished in oil. Fig. 185 shows a convenient way to dip cutters. The washer should not be any larger than necessary to hold the cutter, for it hinders the cooling where it touches the cutter. The temper is drawn by heating a round bar that is slightly smaller than the hole in the cutter, slipping the previously polished cutter on to the bar and revolving it slowly till it is drawn to the proper color at the teeth, usually a straw or brown.

Hammer. — A hammer should be hard on the face and pene, and tough through the eye portion. The eye can be kept soft by packing it in clay or wrapping it in asbestos yarn to prevent its heating to a hardening temperature.

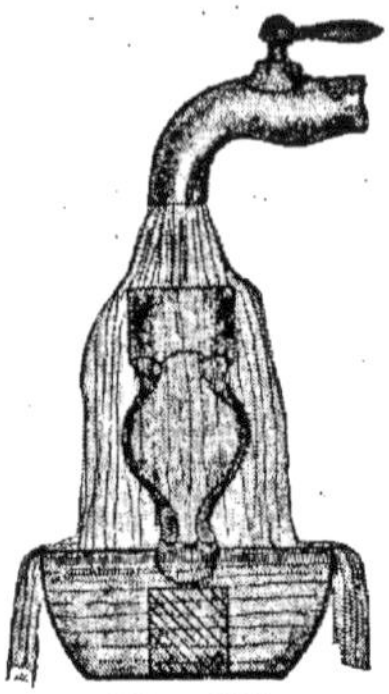

Fig. 186

The author's method is to heat the hammer uniformly all over to the hardening heat, and in place of dipping the whole hammer the pene is held half submerged in a cup of water, while a stream of cold water plays on the face till the whole hammer is cold (Fig. 186). This will leave the hammer in ideal condition, for there is no drawing of temper to be done and before heating there is no bother in covering the parts required to be soft. There should be something in the bottom of the cup for the pene to rest on so that the center of the ball will be flush with the edge of the cup.

Fig. 187

A hammer can be hardened by dipping it all over, and the temper drawn at the eye by placing in the eye a hot iron the shape of the eye. This method is used when the colors are desired for show.

Fig. 188

Thread Cutting Dies. — Rectangular dies made in halves should be heated in a muffle and hardened in water or oil. The temper is drawn, after polishing (Fig. 187).

Spring Dies are tempered as shown by (Fig. 188).

Solid, Round, or Square Dies are hardened in the same way as other dies. The temper can be drawn best by placing the die in a heated iron frame (Fig. 189); this will allow the heat to flow equally from the periphery of the die toward the center, till the proper color is reached at the cutting teeth. Care must be taken that the frame is large enough to fit the dies loosely when cool so that it will not contract and grip the die as it cools.

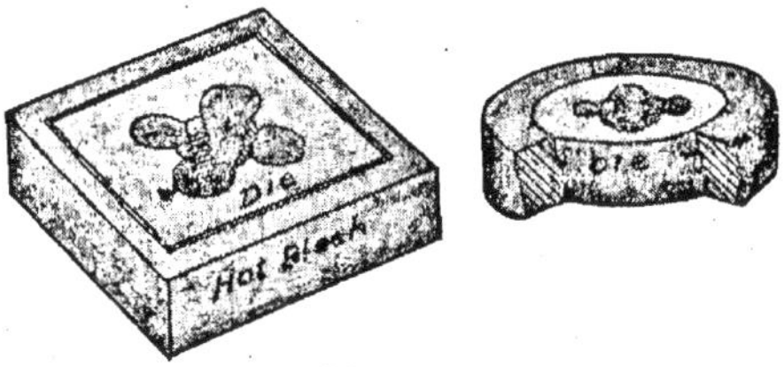

Fig. 189

Counter Bores (Fig. 190) having a deep center hole as at (*a*) and other articles having holes that do not need to be hardened are heated for hardening and tempering after filling the holes with clay. The counterbore should be heated to the lowest heat that will cause it to harden and then it should be dipped into lukewarm water. The temper can be drawn from the shank as in class 1.

Fig. 190

Ring Gages. — If they can be hardened all over and ground to size and shape[1] the pieces can be dipped in water with the hole directly over or under a faucet so that a stream of water can be forced through the hole and the temper drawn by moving it in heated sand. If the hole only requires hardening, Fig. 191, taken from "The American Steel Worker," shows clearly a most excellent

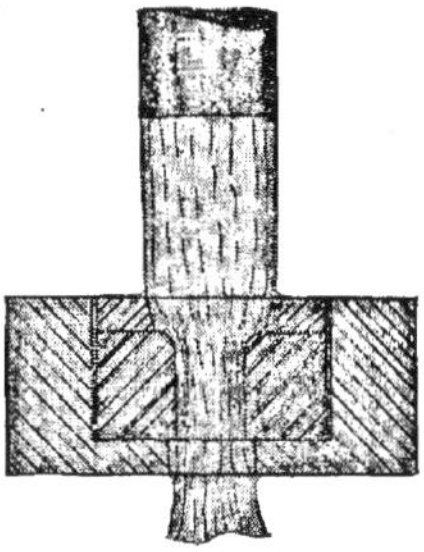
Fig. 191

[1] Round pieces have a tendency to become elliptical.

method. Here a stream of water is forced through the hole in the gage, cooling and hardening the sides. The gage is protected from the effects of the water at all other parts by the surrounding block and washer; which protected parts cool slowly and remain soft. The bulk of the gage, remaining soft, allows the hole to remain true in size and shape.

Pieces with Holes near One Edge should be dipped slowly and have the hole enter the cooling bath last.

Press Dies or Drop Forge Dies that require the working face to be very hard and the bulk of the block tough to resist shocks should be dipped in water at about 60° F., for a minute or so, and then raised and played upon by a stream of water that will cover the face of the die, until the block is cool. To remove all internal strains it is well to reheat the die in hot water after it is hardened.

Tempering in Oil. — When it is not necessary to show the temper color, articles can be tempered very quickly by dipping them in oil maintained at the required tempering temperature, secured by regulating the fire so that a thermometer placed in the oil will stand at the temperature which will produce the temper desired, say 630° F. if we wish to have the piece tough. The piece is dipped into the oil and held there till it is the same temperature as the bath, when it is withdrawn and cooled in water. Numerous small articles can be placed in a wire basket and together heated in the oil and then cooled.

Thin Articles, like knife blades and slitting saws, that are likely to warp in hardening, are cooled between two heavy blocks of iron. The pieces should be heated by being laid on a hot plate or on the bottom of a muffle. When heated the pieces are placed on one of the blocks which has been previously treated with lard, raw linseed oil, or tallow, and the other plate similarly treated is placed on top as quickly as possible and left until the pieces are cold. Saws should be picked up by the center hole to avoid spoiling the teeth. Special tongs can be made for this purpose.

Springs are first hardened by being plunged into oil or tallow and then the temper is drawn. The most common method of drawing the temper is to burn off the oil that adheres to the spring and then to dip it into water to stop further drawing. In large work it may be necessary to burn oil on it two or three times. This process is known as *flashing*. With springs of unequal section the temper is drawn in oil heated to 560° to 630° F. Care must be taken to prevent the oil from catching fire. Small springs are sometimes covered with charcoal dust and the temper is drawn by burning off this dust.

Pack-Hardening. — By this method articles can be given an extremely hard surface when dipped in oil, with little change in shape and but slight danger of cracking. The method is to pack the articles to be hardened into a mixture of granulated charcoal and charred leather (the use of bone is not advisable on account of the phosphorus it contains), into boxes in a manner similar to that in packing for pack-annealing, heat the box and contents to the refining temperature, and plunge the heated pieces into oil to cool. When the pieces have reached the proper temperature it can be told by wires as described before.

Case-Hardening. — The surface of wrought iron or mild steel can be made exceedingly hard while the interior retains all the strength and toughness of the original piece. This process of making the outside or skin of wrought iron or mild steel into a hardened form of steel is called *case-hardening*. It can be done in *two* ways — by dipping in melted potassium ferrocyanide, or by packing in charcoal or bone.

Potassium Ferrocyanide Method. — This method is used when rapid work is required. It does not produce as deep a coating of steel as the other method, but is convenient and rapid. If much work is to be done the potassium ferrocyanide is melted in an iron pot; the iron is heated to a red, dipped into the melted cyanide, heated again to the refining temperature, and plunged into water. When but one or two pieces are to be hardened, the iron is heated to a bright red, and powdered potassium ferrocyanide is sprinkled on the parts to be hardened. The heat in the iron will cause the cyanide to melt and cover all parts of the piece where hardening is desired. The piece is then reheated to the refining temperature and is dipped into water.

Packing in Charcoal Method. — The pieces to be hardened are packed into an iron box with an equal mixture, by measurement, of granulated wood charcoal and raw bone. The packing is done as described under pack-annealing. Test wires should be used to tell when the pieces are raised to the proper heat (a good red) to absorb the carbon, and the time should be counted from that point. The time the box should be left in the furnace depends upon the material and the size of the work. Pieces $\frac{1}{4}''$ thick should stay about 2 hours. Fair-sized work requiring a hard but not very deep coating can be left 5 hours. As a rule a coating about $\frac{1}{8}''$ deep can be obtained in from 15 to 18 hours. When the pieces have been heated the set length of time, the box is withdrawn from the fire and the pieces plunged into water as quickly as possible. The pieces can be colored by being allowed to fall from the box into the water, a distance of 12 to 16 inches.

Straightening Bent Tools.—Sometimes tools will warp in hardening. They can be straightened by placing the pieces between the centers of a lathe with the bowed side towards the tool-post. A piece of iron should be placed in the post so that it will bear on the tool when the transverse feed is moved forward (Fig. 192). The tool is covered with oil and heated with a flame from a Bunsen burner with a wing top, till the oil begins to smoke; then pressure is applied to the tool with the transverse feed until the tool is bent slightly in the other direction, then removed and cooled.

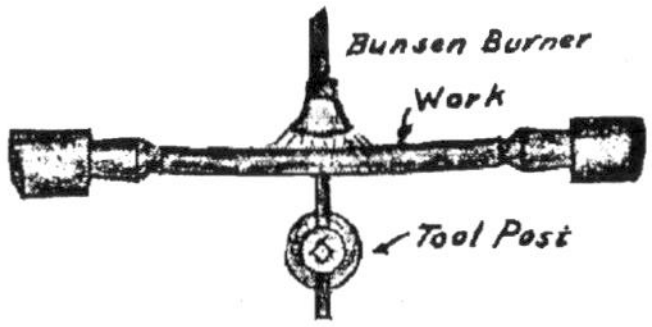

Fig. 192

QUESTIONS FOR REVIEW

How does machinery and tool steel differ from wrought iron? What are the two principal classes of steel? How do they differ when dipped into water at a high heat? What is meant by temper? What is the principal hardening element in tool steel? What is meant by point? Why is heating one of the most important parts of the hardener's duties? What must be looked out for in heating tool steel? Why does charcoal make a good fuel? Why is a muffle furnace best for heating tool steel? Why is an open forge good for heating tool steel when it is afterwards to be turned in a lathe? Why must tool steel be heated uniformly? Name the principal heating baths. Why must they be pure? What effect has sulphur on the steel? What are the disadvantages of the lead baths? The cyanide? What are the advantages of each? Why should the baths be heated so that the fumes cannot enter the room? Describe a furnace used for a cyanide bath. How is the gas torch used for heating small pieces? Where should they be located and how? Why is their location important in obtaining uniform results? What effect has heating on the grain of tool steel? When steel is being hardened why should the lowest possible heat be used? What causes the temperature to vary? What is the decalescence point? What takes place in steel heated above or below this point? Give a method for determining the proper temperature at which to harden a piece of steel. What is drawing temper? Give rules for heating. Why should steel be dipped on a rising temperature and never on a falling temperature? What is reheating? Why is it used? What is annealing? How many ways are there of annealing? Describe each. Give the rules for annealing. Make a diagram showing the effects on the size of grain and the kind of carbon produced by heating steel to different temperatures and slowly and rapidly cooling from these different temperatures. What are hardening baths? Name some. What is the advantage of their use? How are they used? With what class of work are they used? What advantage is oil on the surface of the water in a hardening bath? Why is flowing water sometimes used? What are tempering colors? How are they produced? What do they indicate? How many classes of hardening? Where does the heat for drawing temper come from in each case? Describe tempering in oil. What advantage? How are springs tempered? What is pack-hardening? What is case-hardening? How many methods? How do we case-harden with ferro-potassium cyanide? How is case-hardening done by packing in charcoal? How is bent work straightened?

CHAPTER XIII

HIGH-SPEED TOOL STEEL

Carbon and Self-hardening or Air-hardening Steels. — The difference between ordinary carbon steel and self- or air-hardening steel is as follows: Tools made from the former first are hardened by being heated to redness and then suddenly plunged into water, and then tempered by some method of reheating to the desired tempering point and cooled again; whereas, self- or air-hardening steels acquire a definite degree of hardness whether cooled rapidly or slowly. Hence, tools made from this steel need only to be forged to shape, the temper being obtained by simply allowing them to cool in the air. The property of self-hardening makes tools that are made of this steel able to stand high temperatures while cutting without reducing the hardness of the tool.

High-speed Steel. — It has been discovered recently that steels containing chromium and tungsten, known as *High Speed* steels, when rapidly heated to a white heat and then cooled steadily in a current of air, have their endurance increased wonderfully.

These steels are capable of taking such heavy cuts and at such rapid speed that the tool will actually become red-hot. Herein is the distinction between carbon steel and the high-speed or self-hardening kinds. Tools made from the former, if so heated by working, would become soft and useless by having their cutting edge rapidly worn away, while the high-speed tools similarly heated are unharmed.

High-speed steel when annealed by certain perfected processes can be machined easily so that it can be utilized to make cutters and similar tools. The greatest advantage, however, possessed by this steel over the carbon steel is that there is very little danger of loss in the hardening bath, where so many costly tools are ruined, for it is only necessary to reheat such a steel and cool it in an air blast to cause it to regain the hardness it possessed before annealing.

The Working of High-Speed Tool Steel.[1] — The forging of a high-speed tool should be done at a good yellow heat, 1850° F., and should never be done below a bright red. It is better to reheat the tool several times than to work it below a bright red in one heat. After forging, the point of the tool should be cooled in lime or ashes. The tool should not be plunged directly into the hot fire but should be heated gradually. When the tool is hardened, the nose of the tool is heated slowly, in a muffle, to 1650° F., or to a bright red and then rapidly to 2000° F., or to a white heat. After this the tool is cooled in an air blast, or, if intended for the cutting of soft materials, it may be cooled slowly by being set in a dry place. Then the tool, after grinding, will be ready for use.

Annealing. — To anneal the steel it is heated in a muffle to a temperature of from about 1300° to 1500° F. and kept in the muffle at this temperature for two hours; then it is slowly cooled in ashes.

Grinding. — The way in which tools are ground is of considerable importance, for if not properly followed it may injure the tool permanently by causing it to crack, etc. The best and soundest steels are often ruined in this way.

[1] Most high-speed steels require special methods of treatment that are best obtained from the manufacturer.

High-speed steels should be ground on a well-selected wet sand stone and the pressure should be produced by hand. If tools must be ground on an emery-wheel it is best to grind them roughly to shape before hardening. When this is done they will require very little grinding after hardening, which can be done with slight frictional heating so that the temper will not be drawn in any way, or the cutting efficiency impaired. When grinding tools on a wet emery-wheel, if much pressure is applied, the heat generated by friction will heat the tool to such a degree that the water playing on the steel will cause it to crack.

Hardening and Tempering Specially Formed Tools. — When such tools as milling cutters, taps, screw-cutting dies, reamers, and other tools which do not permit of being ground to shape after hardening, are made from high-speed steel, they must be hardened and tempered, as follows: They should be heated in a specially arranged muffle furnace which consists of two chambers lined with fire-clay. The furnace should be gas, or oil-fired and so constructed that the gas and air enter through a series of burners at the back to produce a temperature of 2200° F. that may be steadily maintained in the lower chamber, while the upper chamber can be kept at a much lower temperature.

The operation is as follows: The tools are first placed upon the top of the furnace until they become warmed through, placed into the upper chamber and uniformly heated to a temperature of 1500° F. (a bright red heat), and then placed into the lower chambers, where they remain until heated to 2200° F., or till the cutting edges show a bright yellow heat, at which temperature the surface appears glazed or greasy. The cutters, while the edges are still sharp and uninjured, are withdrawn and revolved in an air blast until the red has disappeared. When the cutter has cooled to the point that will just permit it to be handled, it should be plunged into tallow heated to 200° F. The temperature of the tallow is then raised to 520° F., when the cutter is removed and plunged into cold oil. If the cutter is large, it can be allowed to cool to the normal temperature in the tallow. If an air blast is not available, small cutters may be hardened by being plunged into oil from the yellow heat.

Another very good method of tempering is by means of a specially arranged gas and air stove. The articles to be tempered are placed in the stove and heated to a temperature of 500 to 600° F. Then the gas is shut off and the furnace and contents allowed to cool slowly.

It is highly important that the initial heating be done slowly and thoroughly or the pieces are likely to be spoiled by warping or cracking due to unequal expansion.

QUESTIONS FOR REVIEW

What is carbon steel? What is air hardening steel? What is high speed steel? Tell how each differs. Tell how to harden and temper tools made from high speed steel. Describe the working of high speed steel in the forge fire. Describe the annealing of high speed steel. Describe the grinding of high speed steel.

CHAPTER XIV

ART IRON-WORK

ORNAMENTAL or art iron-work is so varied that it is impossible in a short chapter to take up more than a

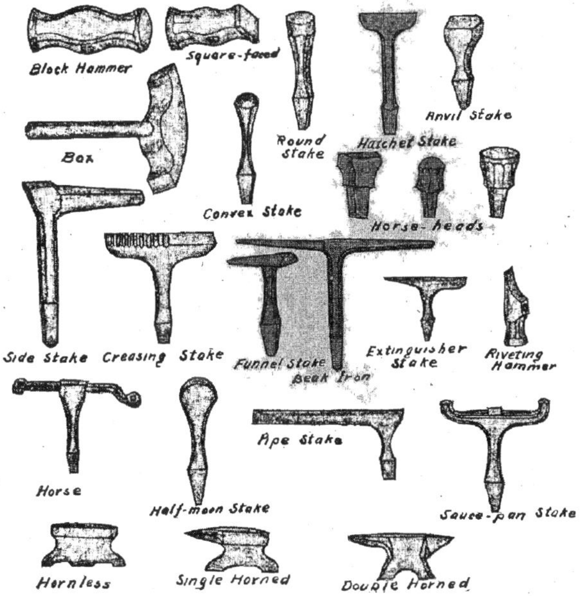

Fig. 193

few of the fundamentals. Therefore, only the more commonly recurring details will be described.

Tools. — The tools used in art iron-work include all of the forge-shop tools, various small anvils, stakes and hammers (Fig. 193), small hand and tail vises (Fig. 194), flat and round nose pliers (Fig. 195), a large assortment of chisels (Fig. 196), and chasing or repoussé tools (Fig.

Fig. 194

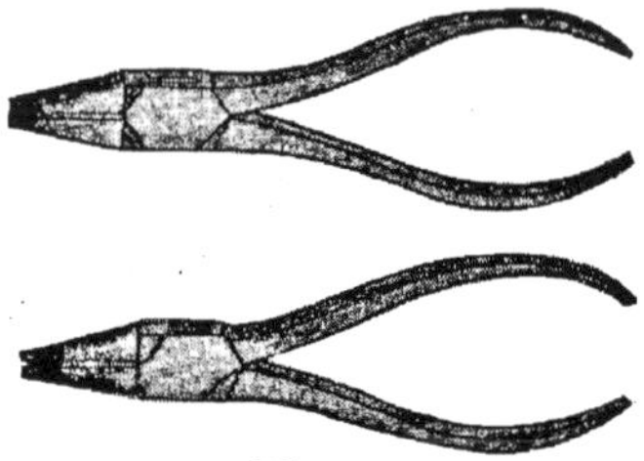

Fig. 195

197). The shop should be supplied with files of all sizes and shapes; such as flat, triangular, square, round, knife, half round, and entering or cross files (elliptical). Taps, dies, reamers, drills, and broaches are also much used.

OPERATIONS

All of the regular forging operations lend themselves to art smithing; such as flattening, upsetting, drawing out, and welding. The fuller and especially the swage find much application, while bending and welding are the principal parts of the work.

Embossing, or punching out bosses (rounded bumps), is done in two ways. The metal is driven out with swages while hot or on thin work while cold, by resting it on wood or lead and using the pene of the hammer. Large bosses or saucer-shaped projections are hammered out cold, in wood, lead, or pitch when the metal is thin, and when thick the metal is heated and the work done on the swage-block. This is accomplished with the pene of the hammer, by starting at the center and working toward the outside to stretch the metal and force it into the depression. By moving and turning the work, the desired shape can be made.

Spinning or Impressing. — Thin sheets are pressed by means of a blunt or a rounded end tool into a form that is rapidly spun on a lathe.

Chasing and Engraving.—Chasing is forming a design on thin metal, such as the veining of leaves, by the depression of the surface with dull, chisel-shaped tools (Fig. 197). Engraving is a similar operation wherein the design is cut into the metal with small chisels or gravers.

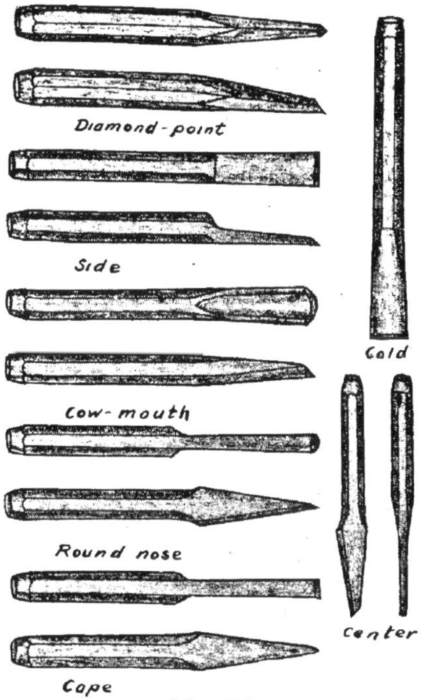

Fig. 196

Etching is the formation of a design in the metal by the eating away of the surface with an acid, — usually sulphuric or nitric. The metal that is to remain unetched is covered with asphaltum paint and the whole is placed in the acid until the design is eaten to the desired depth. The paint can be removed by means of gasoline or benzine.

METHODS OF JOINING

Welding, Brazing, Hard Soldering, Riveting, and Screwing are the common ways of joining work in art smithing. Welding and brazing are the best, but for the novice the most difficult. The rivet makes a good fastening, but it is not always possible to drive or head it. When a screw is used there are two methods of using it: first, to have a nut and join the pieces in the ordinary way (*a* Fig. 198); or second, to tap the inner piece and screw the bolt into it in place of the nut (*b* Fig. 198).

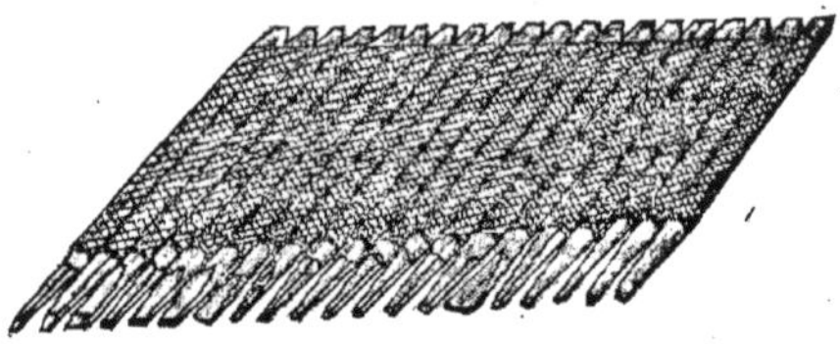

Fig. 197

In both riveting and screwing the pieces can be prepared in several ways (Fig. 199). At (*a*) is shown a method of joining in which one of the pieces is drawn down to a thin edge. At (*b*) a steplike cut is made equal in depth to the piece that is to be joined; (*c*) shows a method of crossing two pieces by offsetting one or both of the pieces where they cross; (*d*) shows two pieces crossed where each is cut in half of the thickness of the piece in the same way as in a half-lapped joint in woodwork.

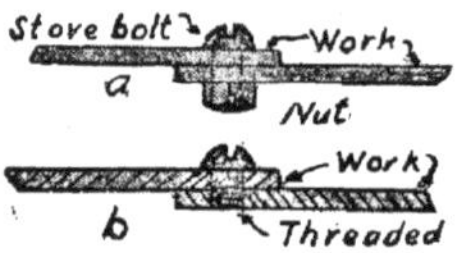

Fig. 198

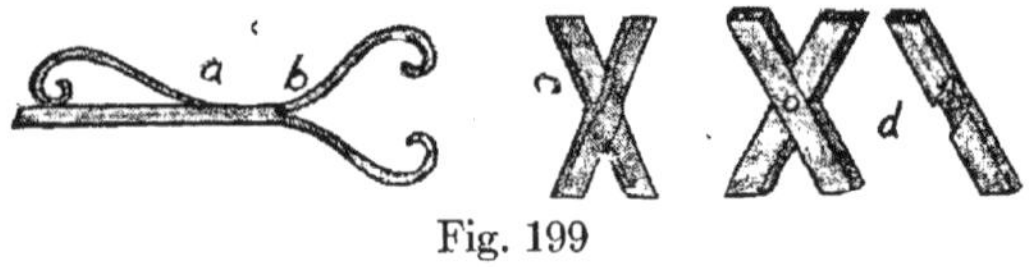

Fig. 199

Fig. 200 represents a method of fitting wherein one bar is passed through another, the whole being either punched or drilled, though usually punched so as to spread the piece as at (*a*) and (*b*). The shape of the bulged or spread part shown at (*a*) would be finished in a V-shaped swage. A method called *tenoning* shown at (*c*) is used much in fastening cast or turned ornamental pieces to the ends of the bars. Collars (Fig. 201) are very commonly used to hold two or more pieces together when they are securely fastened at some other point. If the collars are shrunk on, a much tighter and better joint is obtained. This is done by heating the collar to a red heat and quickly placing it in its proper position, where, upon cooling, it will shrink and grip the work very tightly.

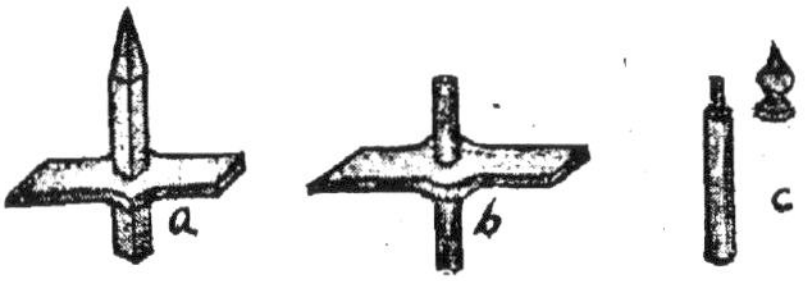

Fig. 200

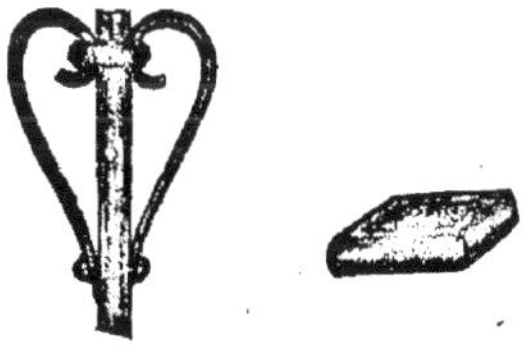
Fig. 201

The Wedge is occasionally used in art iron-work to hold or draw pieces up tight.

Folding (Fig. 202) is used to join thin sheet metal. There are three types: (*a*) the single fold; (*b*) the overlapping fold; (*c*) the double fold. Pieces are folded and placed in position as shown at *a*, *b*, *c*, and closed to a tight joint by means of a hammer or a machine called a seamer.

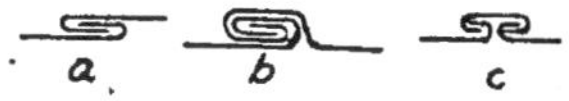

Fig. 202

METHODS OF MAKING THE MORE COMMONLY OCCURRING DETAILS

In Fig. 203 the lines (*a*) are cut with either a chisel or a chasing tool; the small dots (*b*), with sharp-pointed punch; the large ones (*c*), with a blunt punch; and the circle (*d*) either with gouge-shaped chisel or with a round, hollow punch. Cutting away the edges as at (*e*) is called *fretting* and can be done with either cold-chisel or file.

Fig. 204 shows work produced by the use of top and bottom swage. The work is upset at the point where the ornament is to be made, or else a proper sized collar is welded on and the ornament molded to shape by hammering the metal between swages. In later practice such ornaments are cast or drop-forged and bored out, to fit tight when placed on the bar, when the ends of two separate bars are each passed halfway through the piece and fastened by means of pins.

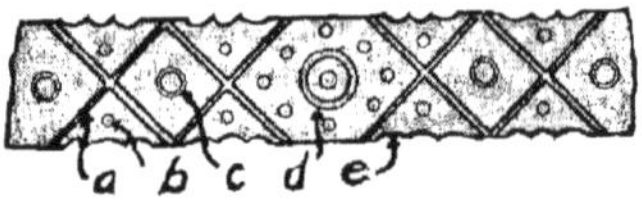

Fig. 203

Fig. 204

Twisting is one of the most commonly used details in this type of work. Thin pieces are usually twisted cold while the larger pieces must be heated. Fig. 205 shows the method of making a twist. The piece is gripped in a vise at the place that is to be one end of the twist, and turned by a wrench at the other end the proper number of times. When possible it is well to have a special wrench (Fig. 206), which fits the shape and size of the stock to be bent. With such wrench a more even pressure can be applied and bending the pieces avoided. When using an ordinary wrench it should be backed up with the

Fig. 205

Fig. 206

left hand. Straight pieces can be prevented from bending, while being twisted, by being placed inside a piece of gas-pipe, which is the same length as the desired twist.

The scroll is made as follows: The end is bent to the arc of a circle over the horn of the anvil (*a*) (Fig. 207), then placed on top of the anvil (*b*) and struck hammer

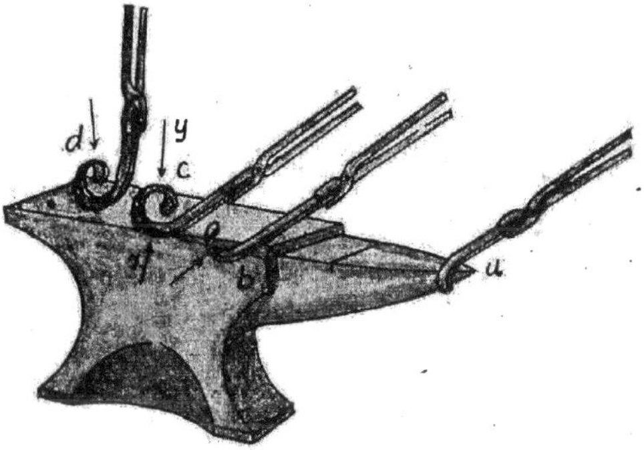

Fig. 207

blows in the direction of the arrow, the piece being moved forward continually, which cause the scroll to form as at (*c*). If the scroll bends too fast, that is, closes too rapidly, the hammer should strike nearer the anvil as indicated by arrow (*x*), or the end of the piece held in the tongs should be slightly lowered. If the bend is not rapid enough, the hammer blows must be higher up, as indicated by arrow (*y*) or the piece must be raised or rolled on the face of the anvil to make the scroll rest on the anvil (*d*), when it can be hammered as indicated by the arrow.

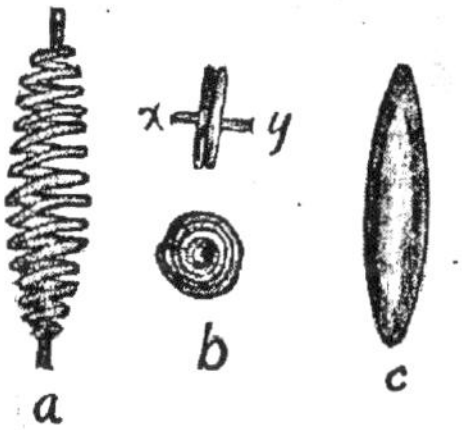

Fig. 208

When many scrolls just alike are wanted it is best to make a former of heavy iron and bend the scrolls on it.

The Spindle Shaped Spiral Twist (*a*) (Fig. 208) is made from round rods or large wire, by two methods.

One is to coil the wire as in (*b*), and then pull out the scrolls to the proper shape by pulling the ends (*x*) and (*y*). The other way is to turn a piece of wood (*c*) to the proper shape, though slightly smaller than the desired size, and bend the wire around this, and afterward burn out of the spiral the wood.

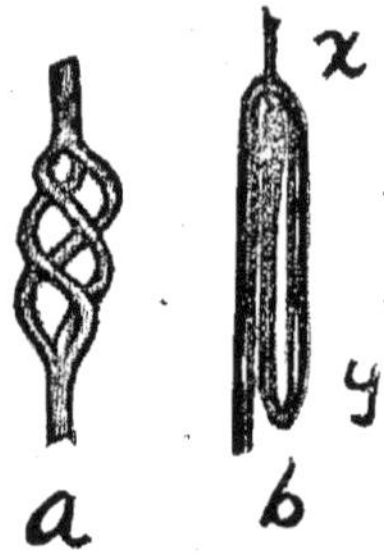

Fig. 209

The shape (*a*) (Fig. 209) is made by folding the stock, shown at (*b*), as many times as there are to be branches and then welding each end, (*x*) and (*y*). The folds are then heated and given a slight twist, and at the same time the ends are pushed towards each other to cause the branches to spread apart.

Interlacings (*a*) (Fig. 210) are made by making two loops in the form of a figure 8 and threading the ends of the stock through the loops, in a manner similar to the tying of a knot. In (*b*) the figure "8" is first made, and then the parts (*x*) are shaped, the ends (*y*) are drawn through the loops, ends (*z*) welded together, and the ends (*y*) welded to the work.

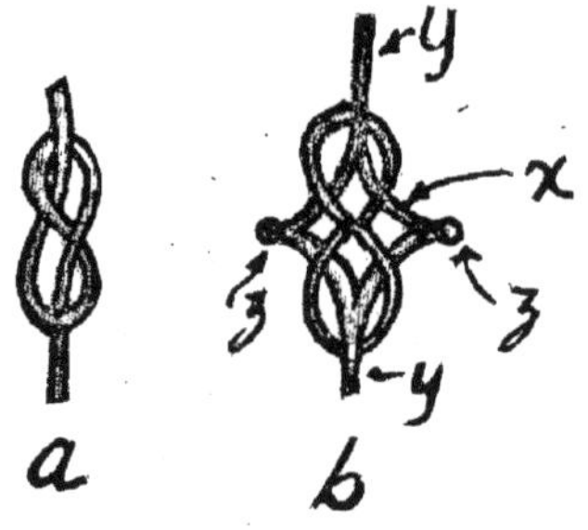

Fig. 210

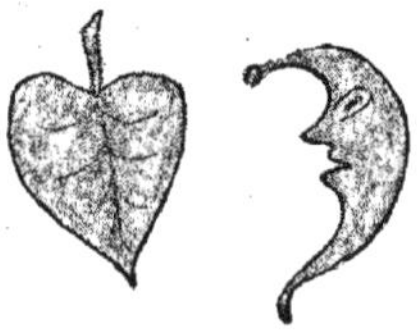
Fig. 211

Leaves and Ornaments (Fig. 211) are cut from sheet metal with shears or by other means, and are then veined with a chasing tool or chisel and crinkled to the desired shape by means of flat or round nose pliers. When many identical ornaments are wanted, dies are often made to cut them from the metal by a single blow. Often the ends of bars are flattened into thin sheets and cut to the form of leaves with saw, chisel, and file.

GENERAL PROCEDURE

When a piece of work is to be made, as, for instance, that shown in Fig. 212, the first step is to make a full-sized drawing preferably on a board so that the measurements of the parts can be scaled therefrom and the pieces being made placed thereupon for comparison. The twists (*a*) should be made, the scrolls, (*b*) and (*c*), bent; and the holes (*d*) for screws should next be drilled. The bowl (not shown) should be made next, and if it is to be ornamented with leaves they should be cut out, filed to the proper shape, veined, crinkled, and drilled for riveting or scarfed for welding. Swages should be made for such parts as (*e*) and (*f*). The spiral twists should be made and welded on. When the parts are all completed they should be joined by riveting, the clips (*g*) shrunk on and the whole straightened and given the final adjustment. The piece should now be given a coat of dull black paint.

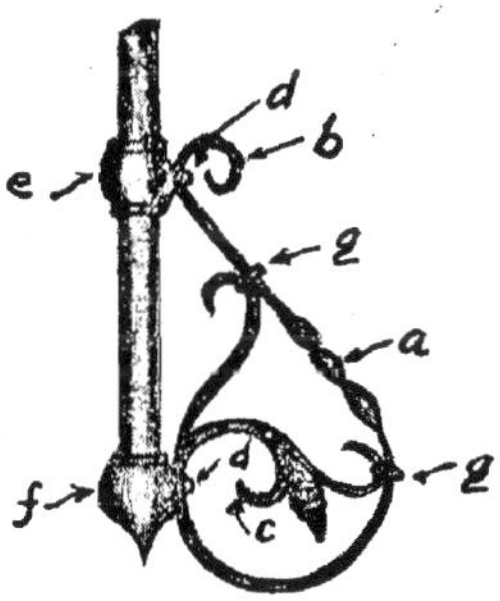

Fig. 212

QUESTIONS FOR REVIEW

What tools are used in art iron work that are not used in ordinary forge work? Does art work call for different operations from those called for in ordinary forging? What are embossing, spinning or impressing, chasing, engraving, and etching? What are the methods of joining? Name some of the reoccurring details found in art iron work. Describe the operation of twisting. Describe the making of a scroll, a spindle-shaped spiral twist, leaves, and other ornaments, and the process of interlacing.

CHAPTER XV

STEAM AND POWER HAMMERS

Power hammers have made it possible to do much heavier forging than otherwise could be accomplished, and have enabled us to make small and medium sized work much more cheaply.

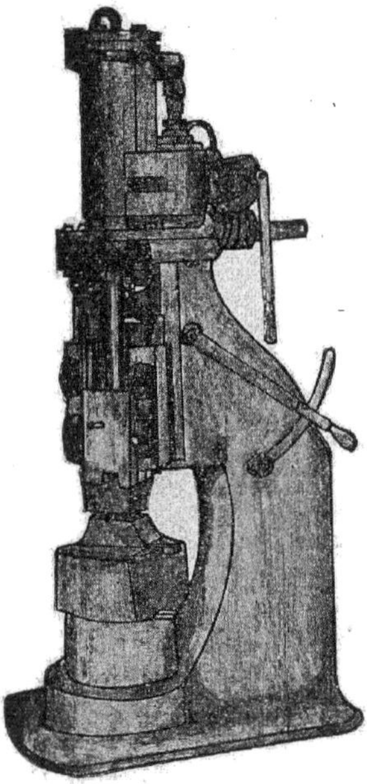

Fig. 213

Power hammers can be divided into two general classes: — Those having a piston acted upon by some expansive gas, and those driven by belts or linkage methods of transmitting power. The first class is represented by the steam and pneumatic hammers, the second in one way by direct-acting hammers, having the hammer head attached to a connecting-rod that connects direct to an eccentric, and in another way by the helve hammers in which the connecting-rod from the eccentric is attached to a beam which is in reality a huge hammer handle.

Fig. 213 shows a single frame steam hammer suitable for all ordinary smith's work. With this hammer the blow can be regulated with the utmost nicety both as to speed and force. It can be stopped and started instantly, and can be made to deliver either a succession of rapid blows or a slow single thud. The size of steam hammers is rated by the weight of the moving parts; i.e., the tup, piston, and piston-rod. Thus a 1000-lb.

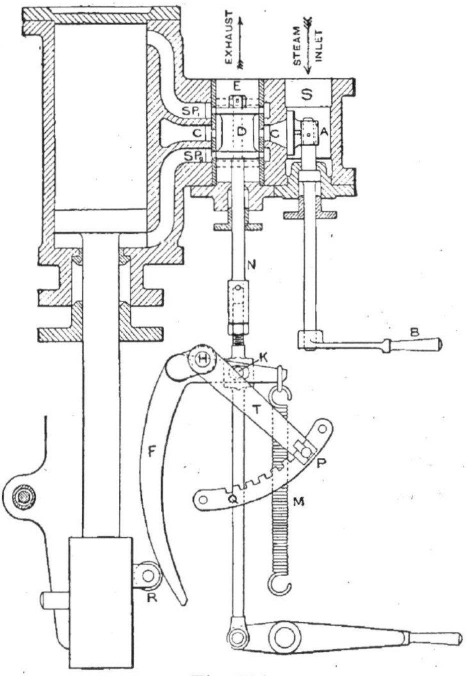

Fig. 214

hammer is one in which these parts weigh 1000 lbs. In rating hammers no account is taken of the steam on the top of the piston. Fig. 214[1] shows a section of the cylinder of a hammer with a combined self-acting and hand-operated valve gear. With the self-acting gear the hammer can be made to give continuous blows, light or

[1] From Nelson's Loose-Leaf Encyclopedia.

heavy, and quick or slow, as long as the steam is on. A single dead blow can be struck at any time with the hand-operated gear. The anvil blocks of steam hammers pass through the base and rest upon pine blocks on a concrete foundation. The hammer in Fig. 213 has a flat face along its piston to prevent it turning around.

Operations. — Steam enters the chest at "*S*" (Fig. 214) and is admitted to the regulating valve port (*C*) by the opening of the starting valve (*A*). This is accomplished by a horizontal movement of the handle (*B*). The regulating valve (*D*) is of the piston type, and by moving this valve up or down the steam entering at port (*C*) can be made to flow either to the upper or lower steam port ($S.P_1$) or ($S.P_2$) respectively. The dotted lines show the extreme travel of the valve. When the steam enters through the port ($S.P_2$), the piston carrying the hammer moves up and the steam above the piston exhausts through port ($S.P_1$) and out at (*E*). On the reverse stroke the travel of the steam is from (*C*) around (*D*) and through ($S.P_1$) to the upper part of the cylinder.

The exhaust is out through ($S.P_2$) and up through the piston valve (*V*), which for this purpose is hollow, to the exhaust pipe. The hammer is worked by hand by moving the handle (*L*) up or down. This brings the valve (*D*) in position to admit the steam either above or below the piston and the piston and hammer descend or rise under the action of the steam. To strike a dead blow the lever is pressed down. (*L*) is not used when the hammer is self-acting. The hammer is made self-acting by causing the curved lever to work about the movable fulcrum (*H*) kept in contact with the roller (*R*) on the hammer head by the spring (*M*), which is attached to the short arm of the lever and the frame of the hammer. As the piston ascends and draws up the head, the lever (*F*) is moved to the right and the valve stem (N) is lifted by the movement of the short lever arm so that the valve (D) is in position to allow steam to enter the upper port ($S.P_1$) and bring the piston down again. As the piston and head move down the spring forces the lever (F) to the left, causing the valve (D) to descend. This allows steam to enter ($S.P_2$) to cause the piston to rise. The length of the piston stroke is regulated by changing the position of the fulcrum (H) through the lever (T) — position (P) giving the longest stroke and (Q) the shortest.

Compressed air can be used in hammers like the above in place of steam, and in this case they are called *pneumatic*.

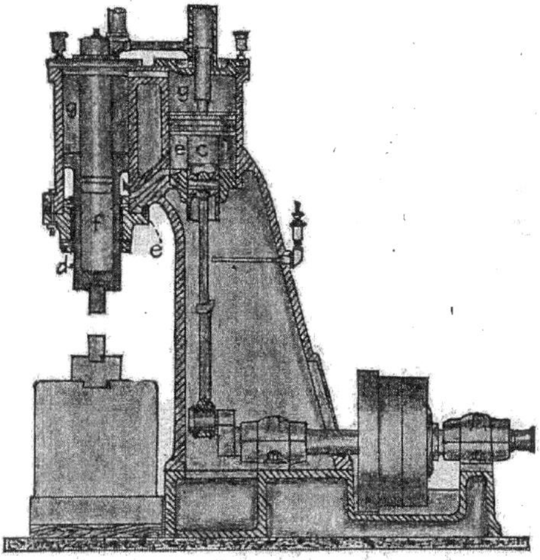

Fig. 215

Fig. 215 shows a section of the Bechi pneumatic hammer which acts as follows: By the lowering of the compression piston (c), the ram (d) is forced upwards by the injection of compressed air into the ram cylinder (e) underneath the top of the ram and by the vacuum in chamber (g) above the ram, produced by the descending piston (c). Thus there are two forces acting jointly to raise the ram. As the piston (c) ascends, the ram (d) is forced down by the flow of the compressed air into the ram cylinder (e) above the ram and the suction in the ram cylinder (e) underneath the top of the ram caused by the ascending of the piston (c). These two forces are augmented by the compression in the inner ram chamber (*f*) which is created by the force of the ascending ram. Thus three downward forces act when a blow is struck.

Fig. 216 shows a Beaudry hammer which will illustrate the type in which the connecting-rod connects the hammer and eccentric. The operation of this hammer is very simple. By pressure of the foot on the treadle at the bottom an idler pulley is caused to tighten the belt on the driving pulley. This drives the eccentric which operates the hammer. The force and speed of the blow is regulated by the pressure of the foot on the treadle.

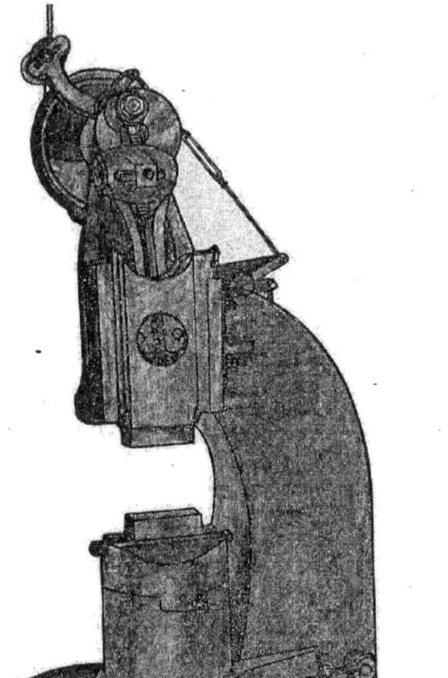
Fig. 216

Fig. 217 shows a helve type hammer. No description of this hammer is necessary more than to say that it operates by foot treadle to throw in a friction clutch. The tighter the clutch is held in, the faster and harder the hammer will strike.

Foundations. — All power hammers should be set on concrete foundations. The anvils on all except the very smallest are separate from the hammer frame and should be mounted on a separate foundation, so that heavy blows will not injure the frame. The anvil foundation should be capped with heavy timbers to give the anvil an elastic support.

Tools. — There are only a few tools used with a power hammer and they are very simple. Tongs should grip the work solidly; for the flat work the box tongs (Figs. 40 and 41) should be used, and for round work those in Figs. 43 and 45. The chisels for power hammer work differ from those for hand work. Fig. 218 (*a*) shows a hot chisel, (*b*) one for nicking cold stock, (*c*) one for cutting square corners, and (*d*) one for cutting round corners. The blades are made of tool steel while the handles are either tool steel drawn out of the same piece as the blade or are wrought iron welded on. As it is not always possible to hold the chisels in just the right position in order to save the hands from jar and prevent breaking handles, the handles are made somewhat thinner at "*x*." This will allow the handle to spring so that the blade can adjust itself. The cutting edge of the chisels are left blunt as shown in the section at "*a*" (Fig. 219). They should never be sharpened as in "*b*" since the force of the hammer blow is very great. The shape for general work is shown at "*a*," while "*c*" and "*d*" show edges

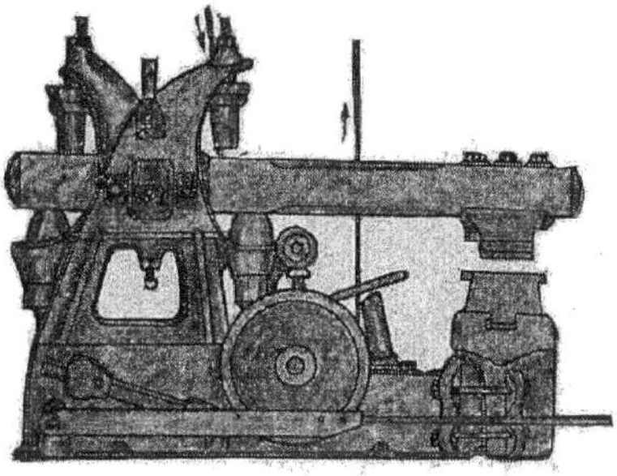

Fig. 217

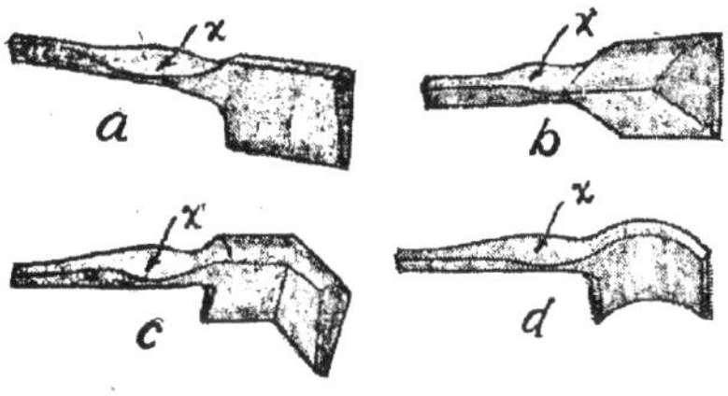

Fig. 218

ground for special work. Round bars of iron or steel usually take the place of fullers. Fig. 220 shows a type of fuller used for some classes of work. The swages used up to 4″ are generally of the spring type (Fig. 63). For larger sizes the bottom swage is like Fig. 221. It is made with three projections one "*a*" to fit the hardie hole in the anvil and the other two to straddle the block. The top swage is held in the hand as in hand forging. Fig. 222 shows a tool for making tapers. Blocks of steel of various sizes and shapes are made use of for bending, offsetting, and the like.

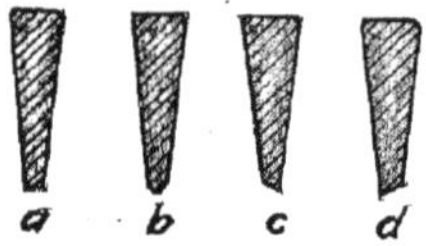

Fig. 219

Fig. 220

Fig. 221

Uses of the various tools. The cold cutter or nicking tool (*b*) (Fig. 218) is made short and stubby to give it strength. Its shape is that of an equilateral triangle or nearly so. With it the cold bars are nicked on two or more sides so that they can be broken. The hot cutter (Fig. 218*a*) can be held either to cut the stock off squarely leaving the end of the piece from which it was cut, bulged in the middle, or it can be held so that both pieces will be bulged in the middle. In either case the cut is started by holding the cutter straight and cutting part way through on all sides. If the cutter is not held straight when starting the cut, as shown at "*a*" (Fig. 223), it will be knocked down as shown at "*b*" (Fig. 224).

Fig. 222

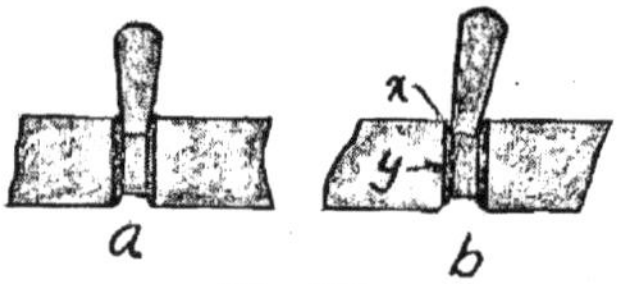

Fig. 223

If it is not desired to have one piece with square end, the cutter can be driven through as started, but if the square end is desired, after cutting all sides, the cutter should be tipped as at (*b*) Fig. 223 and driven through. It can be seen readily that edge "*x*" Fig. 223 will make a square face on portion (*y*). When cutting large pieces the chisel should be driven nearly through the stock as at (*a*) Fig. 225. The piece is then turned over and a piece of rectangular steel placed edgewise directly over the cut "*b*." Then with one blow of the hammer the bar can be driven through and both pieces left with smooth faces, as at "*c*"; whereas if the chisel had been used to make the last cut, the stock would have been as at (*d*).

Fig. 224

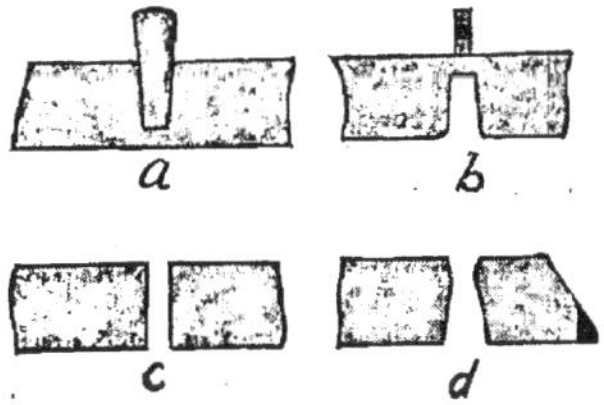

Fig. 225

Fullers. — The ordinary hand fuller, especially the bottom one, finds little use with the hammer. When fullering is to be done requiring top and bottom fullers, two round bars of steel are used (*a*–*a* in Fig. 226). When one essentially straight edge is desired, fullering on one side is done with a top fuller similar to Fig. 220, and if the ordinary shape is desired it is done with a round bar or ordinary fuller with a flexible handle.

Fig. 226

Swages are worked in a manner similar to hand work and, as stated before, up to 4″ the spring swages are usually used. The larger sizes differ from the hand swages only in the method of holding them on the anvil, explained above.

Taper Work. — Owing to the construction of the hammers only parallel sides can be worked without the aid of the special tool (Fig. 222). In Fig. 227 at (*a*) is shown a tool held with the curved side to the work and the piece drawn out and tapered by the fullering action of the curved surface at (*b*). The tool is then turned over so that the flat face levels off the bumps and makes a smooth tapered surface.

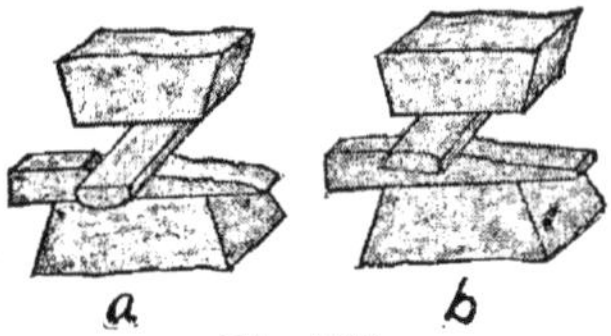

Fig. 227

Bending or Offsetting is accomplished by placing the work between the steel blocks (*a* and *b*, Fig. 228) and hitting a blow with the hammer. The piece will be bent as shown at (*c*).

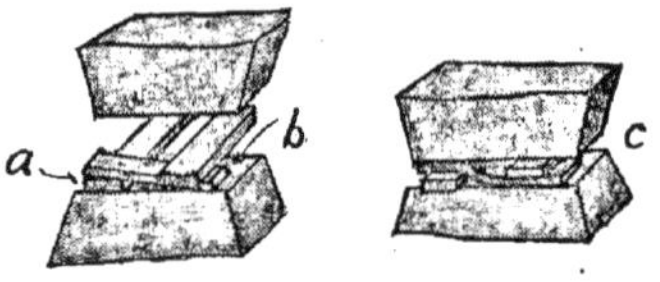

Fig. 228

Drawing Out. — The hammer is the most useful tool for drawing out stock. The same care must be observed in drawing down to round as in performing a similar operation with a hand-hammer. It must be drawn down square first, then octagonal, and so forth until it is round. If a square piece gets out of shape, as (*a*) in Fig. 229, it can be trued up again by turning and striking as at (*b*), rolling it over slightly and hammering it to shape (*c*) and squaring and sharpening the corners by a few blows first on one side and then on the other. Care must be taken to strike a blow that is heavy enough to work the metal all through the cross section of the piece so that the middle will be bulged as at (*a*) in Fig. 230, not hollow as in (*b*) or the

Fig. 229

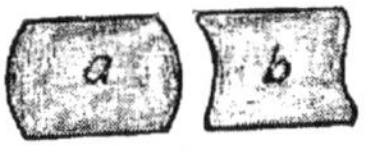

Fig. 230

end of the piece will become cupped and the bar will split. In case of a piece that is too short to straddle the anvil in drawing down in the middle between the two shoulders, the piece is placed on a block that it will straddle and a similar block is placed on top of the work as in Fig. 231. If the piece is to be flat on one side, the flat side can rest directly on the anvil.

Fig. 231

Upsetting. — Short pieces can be upset with the hammer very nicely. Care should be taken to have the blow heavy enough to work the piece throughout.

Press. — On very heavy work the hydraulic press is taking the place of the hammer because a more uniform working of the metal can be effected by it.

QUESTIONS FOR REVIEW

What classes of work are power hammers used on? How many classes of hammers? Name the different types of each class. Make a sketch and describe the operation of the steam hammer. How are the blows varied in force and rapidity? What is a helve hammer? How is a steam hammer rated? What advantage has the Bechi hammer? Describe the chisels for the power hammer. What peculiarity of handle? Why are the anvils set on a separate foundation from the hammer? Why are the foundations capped with heavy timbers? What kind of tongs are used with a power hammer? Why are spring swages used? What is commonly used in place of fullers? Describe the various methods of cutting off stock with power hammers. How is bending and offsetting performed? How is work drawn out under power hammers? Is the power hammer useful for upsetting? Why must stock be given a blow hard enough to work it throughout? Why is the press superseding power hammer work?

CHAPTER XVI

CALCULATIONS

It is often necessary to know exactly the size or length of a piece of stock to be used to make a forging. This information can be best obtained by calculation.

These calculations fall into two classes: Class *A*, in which the length only is to be found or where stock of a known size is simply to be bent; and class *B*, in which the size and section of a piece is changed by drawing out or upsetting. In this case the calculations depend upon the volume.

Class A, *in which bending only takes place.*

The first case in this class will be that of an unwelded ring (Fig. 232). If the outside circumference is figured, the length will be found to be 2.75″ (diam.) × 3.1416 = 8.64″ (circumference = diameter × π) and the inside diameter will be 2″ × 3.1416 = 6.28″. Neither one of these lengths will give a ring of 2″ diameter, for the one will be too large and the other too small, as it has been found that when a piece of iron is bent, the outside stretches and the inside compresses. This being the case, there must be a place that neither stretches nor compresses; this is at the middle of the piece as shown by the dotted circle. If this part of the piece does not change in length in bending, it is evident that the circumference of this circle is the proper length for the stock to make this ring. Thus 2.375″ × 3.1416 = 7.4613″ or $7\frac{15}{32}$″ is the

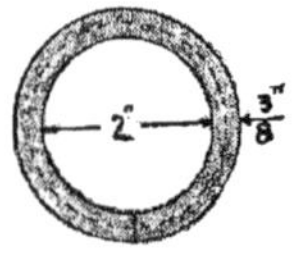

Fig. 232

required length, assuming there is no loss in the fire, and there should be none in the case of bending a ring. In case the ring is to be welded there should be added $\frac{1}{2}$ the diameter of the stock for lapping. The blacksmiths' rule, which gives very close results, is as follows: length of stock for an unwelded ring is $\frac{22}{7}$ × inside diameter + 3 × width of stock.

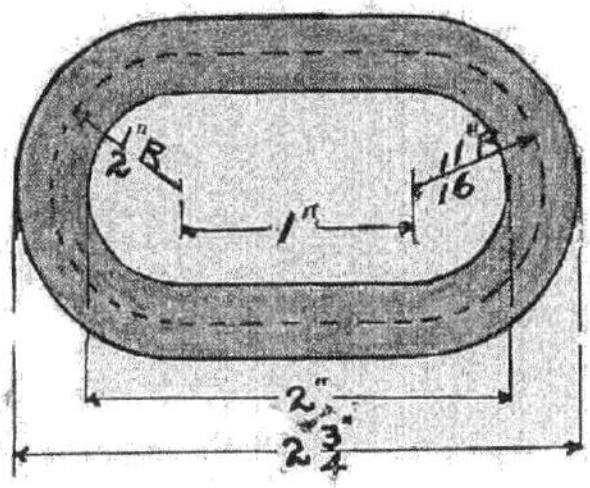

Fig. 233

Chain Link (Fig. 233) gives a slightly different example, as it is a combination of circular ends and straight sides. The two ends together will make a complete circle, the length of which is figured as in the case of the ring above, 1.375″ × 3.1416 = 4.32″. As the sides, being straight, change in no way, by adding twice the distance between the end centers or 2″, the total length is found to be, 4.32″ + 2″ = 6.32″. To this should be added about $\frac{3}{8}$″ for welding.

Arcs of Circles. — The figure 8 (Fig. 234) presents a case almost like the link; only in this case the ends are arcs greater than half a circle. The length of each circular portion is found by the following rule:

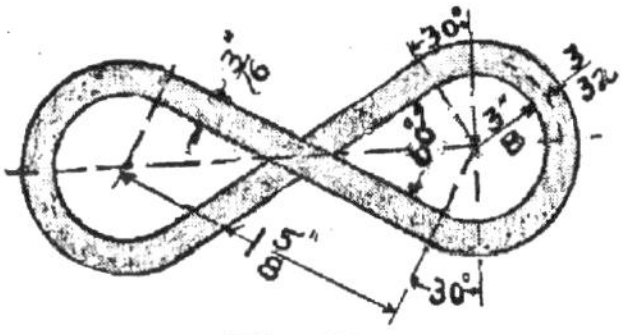

Fig. 234

Number of degrees in the arc × diameter × 0.0087 = circular arc. To this is added twice the distance between points of tangency.

Example. — Degrees in arc 30 + 30 + 180 = 240. Diam. $\frac{15}{16}$ = .9375.

$$2(240 \times .9375 \times .0087) + 2 \times 1\tfrac{5}{8} = 7.16 \text{ or } 7\tfrac{5}{32}.$$

Square Bend. — In the case of the square (Fig. 235), to obtain the length all that is necessary is to add the inside lengths of the legs and the width of the piece the way it is bent or in the case of the figure shown.

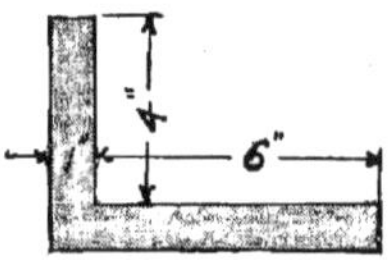

Fig. 235

$$4'' + 6'' + 1'' = 11''.$$

Class B, *in which the section changes in size or shape.*

Drawing from Section of One Size to a Similar Section but of Smaller Size. — For illustration take the problem: How much $\frac{1}{2}''$ square stock will be necessary to produce the piece shown by Fig. 236. The first step is to get the volume of the finished piece which is $\frac{3}{8}'' \times \frac{3}{8}'' \times 6'' = 0.84''$. The volume of $1''$ in length of the $\frac{1}{2}'' \times \frac{1}{2}''$ stock is next determined as $0.5'' \times 0.5'' \times 1'' = 0.25$ cu. in. It will now take as many inches of the $\frac{1}{2}''$ square stock as the number of times 0.25 cu. in. is contained in 0.84 cu. in., or $3\frac{9}{25}''$, or $3\frac{23}{64}''$.

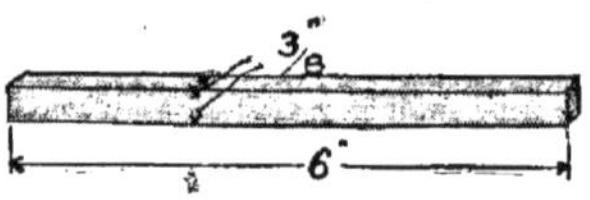

Fig. 236

Fig. 237 is a similar case, only each part, *A*, *B*, *C*, *D*, are considered as separate problems which must each be

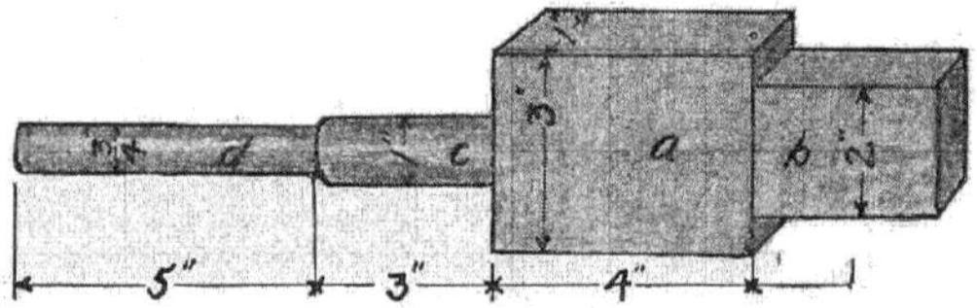

Fig. 237

added for the final result. The exercise is to be made out of $3'' \times 1''$ stock. The problem is how much of this stock is to be taken. For *A*, $4''$ will be needed; for *B*, the length is obtained as above, $3'' \times 2'' \times 1'' = 6$ cu. in. The volume of one inch of length of the stock is $3'' \times$

$1'' \times 1'' = 3$ cu. in., therefore 6 cu.″ ÷ 3 cu.″ = 2, hence 2″ will be needed. The volume of C is $(1'' \times 1'' \times 0.7854'') \times 3'' = 2.356$ cu. in. which, divided by 3 cu. in., gives 0.7854. Therefore 0.7854 inches is needed.

$D = (\frac{3}{4}'' \times \frac{3}{4}'' \times 0.7854) \times 5'' \div 3$ cu. in. $= 0.7363$, or 0.7363″. The total length will be

$$4'' + 2'' + 0.7854'' + 0.7363'' \text{ or } 7\tfrac{17}{32}''.$$

The lengths in these last two examples are for cases in which there are no losses by scaling or otherwise. A small amount determined by experience should be added in each case to make up the loss.

Weight of a Forging. — It is often desired to know the weight of a forging before it is made. This can be obtained by calculating the volume of the piece in cubic inches and multiplying this volume by 0.2779 if of wrought iron and by 0.2936 if of steel. As an example, take the forging of Fig. 237 and consider it as made of steel. The volume of the parts determined as above are $12'' + 6'' + 2.356'' + 2.3'' = 22.656$ cu. in., which multiplied by 0.2936 = 6.7 lbs.

QUESTIONS FOR REVIEW

Why is it necessary to calculate the size of stock or the length of a piece? How many cases are there? What does the second case involve? How are the weights of forgings obtained by calculation? What is the neutral axis?

APPENDIX

A COURSE OF EXERCISES

In this course of study it has been the aim to have each exercise bring out a principle of forgework. In a few cases exercises have been inserted for the practice and skill that it is believed they develop.

Example 1. Drawing out.—Stock, Norway iron $\frac{1}{2}'' \times \frac{1}{2}'' \times 4''$ long.

Explanation. One end of the stock is to be drawn down to $\frac{3}{8}''$ square until it is $5''$ or a little more in

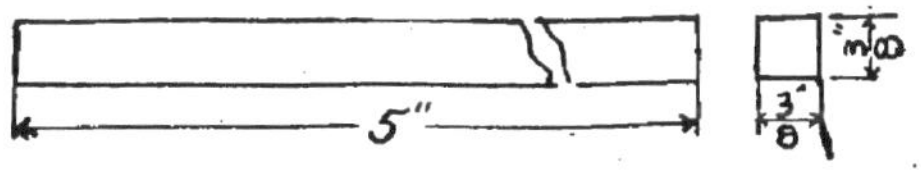

1A.

length. Five inches of this $\frac{3}{8}''$ square portion is then to be cut off on the hardie and both ends trued. The finished piece (Fig. 1 *A*) must be smooth, true to size, square in section and straight.

Operation. Is fully covered in Chapter V, page 62.

Example 2. Upsetting.—Stock, Norway iron $\frac{1}{2}'' \times \frac{1}{2}'' \times 5''$ long.

Explanation. The piece is to be upset to $3''$ in length. The finished exercise (Fig. 2 *A*) must be sound, square, uniform in section, straight and smooth. Determine by calculation the length of the sides x.

Operation. Is fully covered in Chapter V, page 64.

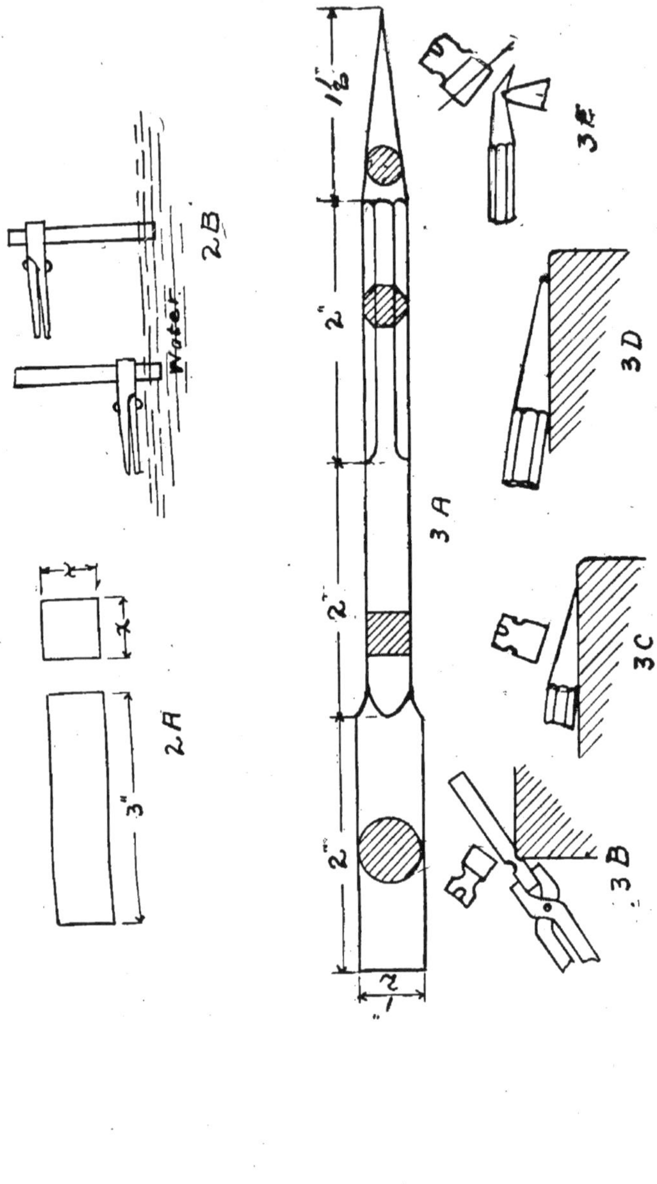
2A
3"
2B
Water
3A
1⅛"
2"
2"
2"
1"
3B
3C
3D
3E

Caution. Have the stock at a white or welding heat. Do not work it after it has cooled below a bright red. Strike squarely on the stock and have it rest squarely on the anvil. Hold tightly in a pair of tongs that fit the piece. If the piece bends, straighten it at once. (Further blows will not upset but will cause it to bend more.) If the ends upset more rapidly than the middle, cool them slightly by rapidly placing first one end in water and then the other. If this is not done rapidly the body of the stock will be cooled. Fig. 2 *B* shows method of cooling.

Example 3. Drawing out to Various Sections. — Stock, Norway Iron $\frac{1}{2}''$ round, 6″ long.

Explanation. The round section is to be drawn down to a square, the square to an octagon, and the octagon to a round point. The completed exercises must agree with Fig. 3 *A*.

Operation. True one end so it will be at right angles with the length of the stock, and make a light center punch mark 2″ from this end. Holding the trued end in the tongs bring the punch mark over the round edge of the anvil and strike one blow. (Fig. 3 *B*). Make a quarter turn and repeat the blow. Continue turning and striking until four fuller marks are made. (Fig. 3 *B*). From these marks draw the stock down to $\frac{3}{8}''$ square. Lay off 4″ from the trued end and mark the four corners as just described for the sides and produce the octagon by hammering down the corners. Again lay off 6″ from the trued end and draw out the point, first to a square (Fig. 3 *C*) and then to a round (Fig. 3 *D*). If there is excess length, cut off as shown in Fig. 3 *E*.

Example 4. Bending. — Stock, Norway Iron $\frac{3}{8}''$ round, 11″ long.

Explanation. Each end of the stock is to be bent so that the piece will form the figure "8" shown in Fig.

4 *A*. The finished exercise must be smooth, true to dimensions, and lie flat.

Operation. True both ends as described for upsetting long pieces, page 65. Lightly center punch at the exact center. Bend the end as described, page 71.

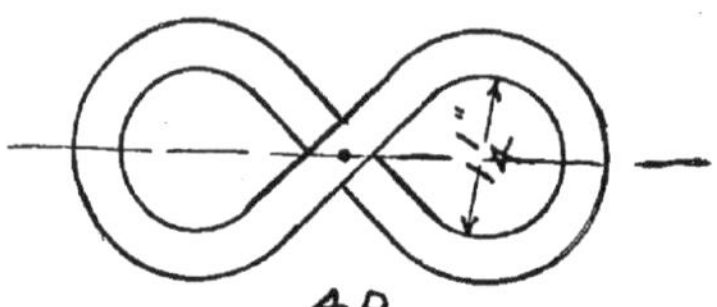

4A.

Caution. As there is little opportunity to smooth the work by hammering, the stock must not be heated hot enough to scale at any time. Never strike the stock directly over the anvil as that will flatten the stock and have no bending effect.

Example 5. Bending to Circle. — Stock, Mild Steel $\frac{3}{8}''$ diameter, $7\frac{3}{8}''$ long.

Explanation. The piece is to be bent to a circle. The finished piece (Fig. 5 *A*) must be free from hammer marks, a true circle and lie flat, and the ends must be parallel where they meet.

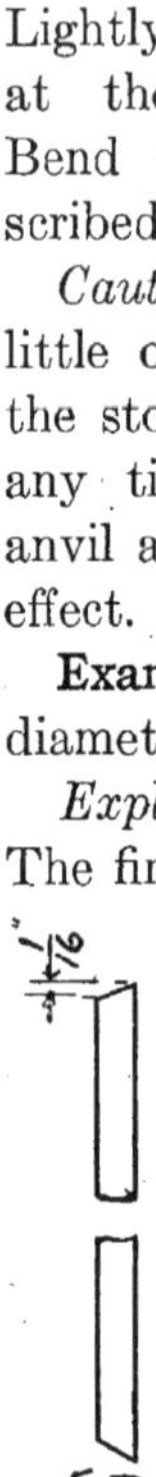

5B.

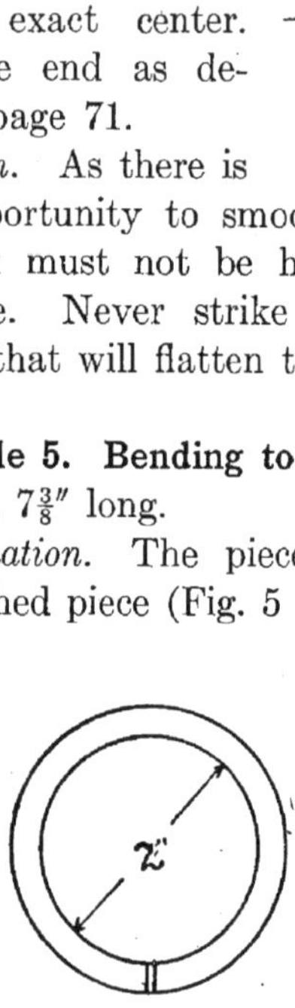

5A

Operation. Bevel the ends as shown at Fig. 5 *B* and bend as directed on page 71.

Example 6. Gate Hook. — (Twisting.) Stock, Mild Steel $\frac{3}{8}''$ diameter, $4\frac{3}{4}''$ long.

Explanation. The finished piece must be smooth and agree with the form and dimensions shown in Fig. 6 *A*.

Operation. Draw down each end of the stock as shown in Fig. 6 *B* and bend as described on page 72.

Caution. In making the bends be careful not to strike the iron directly over the anvil. After the stock has been fastened in the vise for twisting, the work must be done rapidly, for the vise absorbs heat from the stock near it which will cause the twist to be uneven. Keep the stock as straight as possible while twisting.

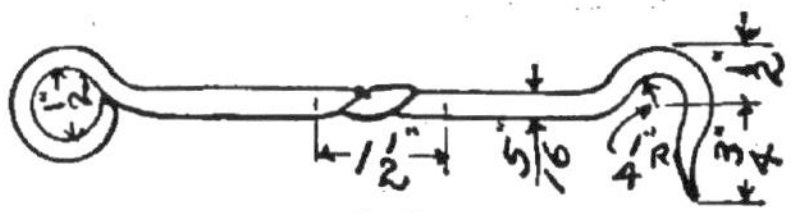

6A

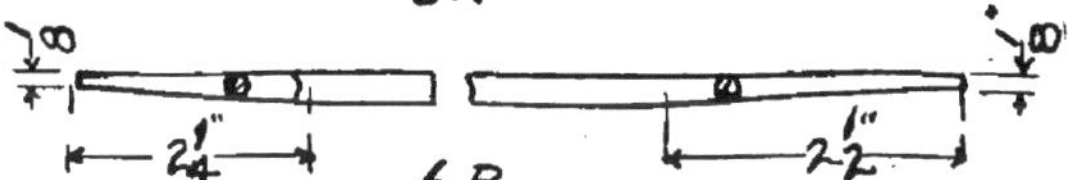

6B

Example 7. Staple. — Stock, Mild Steel $\frac{5}{8}''$ diameter, $3\frac{3}{4}''$ long.

Explanation. The finished piece is to be a well-shaped staple true to the dimensions (Fig. 7 *A*). The points must be even and "in wind."

Operation. Shape the points, and then bend as explained on page 69.

Caution. Care must be used in heating the stock so as not to burn it, as small stock heats very rapidly.

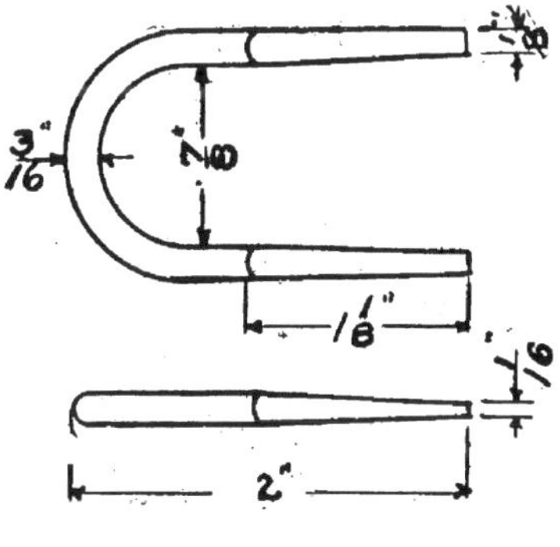

7A.

Example 8. Flat Bend and Punching. — Stock, Mild Steel $\frac{1}{2}'' \times 1 \times 6''$.

Explanation. The stock is to be bent to a right angle and holes for screws punched and counter punched. The piece must be smoothly finished, the legs at right angles and of correct length and all holes true to dimensions (Fig. 8 *A*).

Operation. The piece is bent as explained on pages 68 and 69 and the holes punched as shown on page 79.

Example 9. Fuller Piece. — (Forging for tap wrench.) Stock, Norway Iron $\frac{1}{2}''$ × 1″ × 4″.

Explanation. The stock is to be fullered with top and bottom fullers, the portion between the fuller marks to have the corners rounded, and the arms (*a* and *b*) drawn out to a round section and to the dimensions given in Fig. 9 *A*.

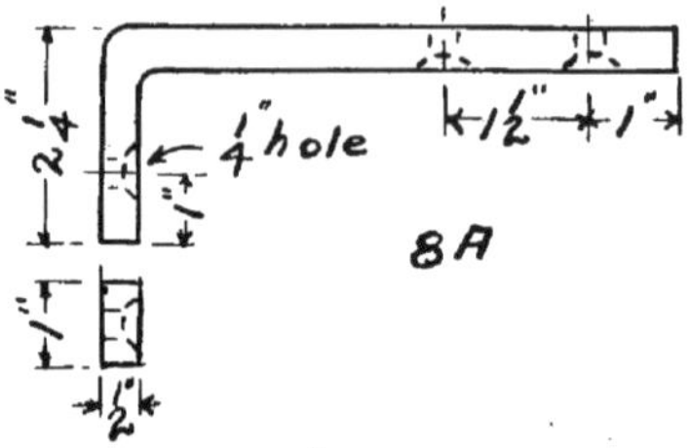

8 A

Operation. The fullering is explained on page 84 and the drawing out on page 62.

Example 10. Door Pull. — (Fuller and set hammer piece.) Stock, Norway Iron $\frac{1}{2}''$ × 1″ × 5″.

Explanation. The stock is fullered 1″ from each end and the part in between drawn down to a round section,

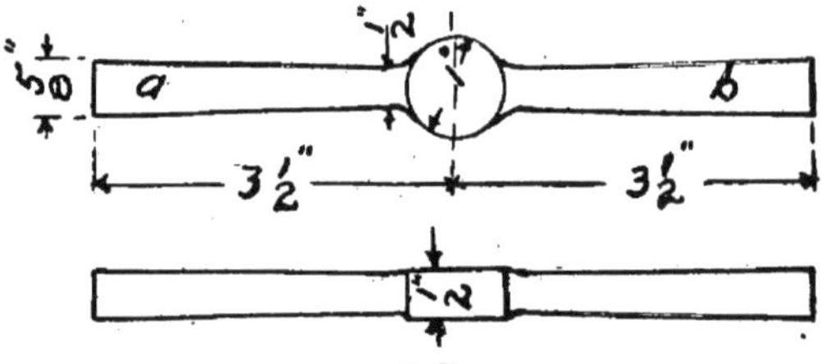

9 A

the ends shaped, holes punched for screws and the center part bent to proper shape. The piece must be true to dimensions (Fig. 10 *A*) and filed to a blue.

Operation. The fullering is done with a top and bottom fuller as explained on page 84. The ends are set down with the set-hammer as explained on page 87 and given the pear shape with the hand-hammer.

Filing to a Blue. Heat the piece to a dull red and rapidly file it all over until the blue (oxide) color appears. Then cool in water.

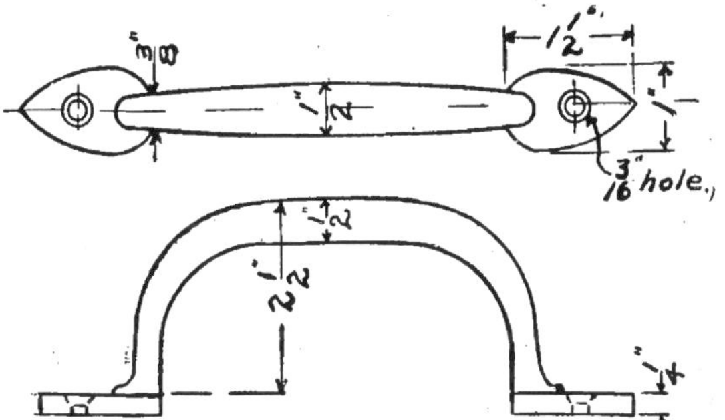

10 A

Example 11. Hammock Hook. — Stock, Norway Iron $\frac{1}{2}'' \times 1'' \times 2\frac{1}{2}''$.

Explanation. A strong well-shaped hook is to be made having a smooth finish and true to dimensions (Fig. 11 *A*).

Operations. The piece is fullered at the middle with top and bottom fuller (page 84) and one end drawn out

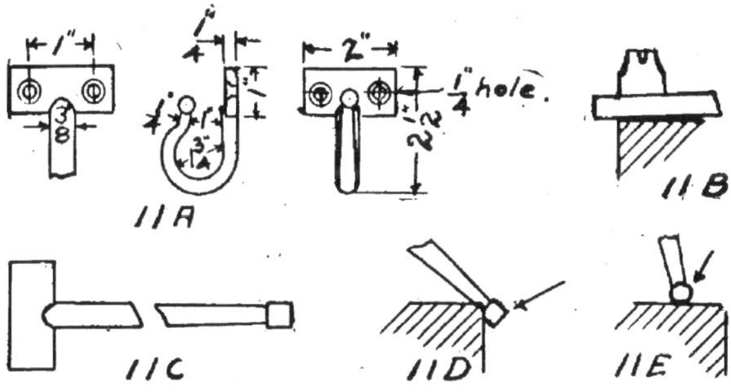

to $\frac{3}{8}''$ round (page 63). The other end is set down and spread out with the set-hammer as explained on page 87. To form the ball, place on the anvil so that the stem projects about $\frac{3}{8}''$ and with a set-hammer, set down as shown in Fig. 11 *B*, to the shape shown in Fig. 11 *C*. The ball is then made as shown in Fig. 11 *D* and 11 *E*. The holes punched and the piece bent to shape.

Example 12. Split Piece. — Stock, Norway Iron $\frac{1}{2}'' \times 1'' \times 3''$.

Explanation. The stock is to be split and opened out to form a fork. The taper must be uniform, the piece smoothly finished and true to dimensions (Fig. 12 *A*).

Operation. The stock should be laid out and fullered (page 84) as shown in Fig. 12 *B* and split as explained on page 77 and the parts drawn out (page 62).

The exercise admits of wide variation in design. Such articles as hooks, oar-locks for boats, and spurs are suggested. The number and length of the tines can be increased and a pitch fork or rake be made.

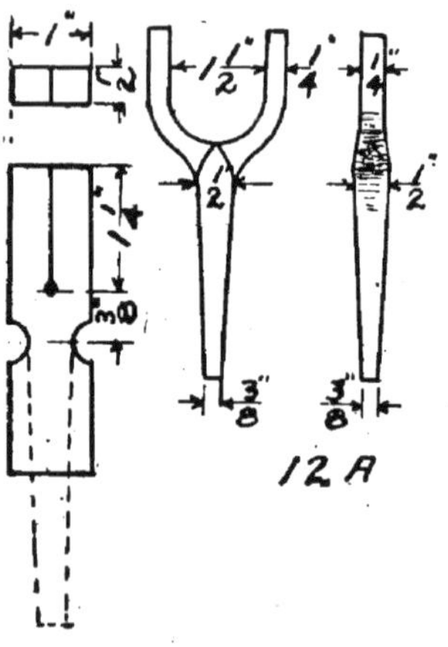

12 A

Example 13. Weldless Ring. — Stock, Norway Iron $\frac{1}{2}'' \times 1'' \times 3''$.

Explanation. The stock is to be split and opened to form a ring, true to dimensions (Fig. 13 *A*), smooth and free from cracks.

Operation. Upset the ends so they will be wider but no thicker than the original stock (Fig. 13 *B*) and round the ends as shown by dotted lines. If the piece is held on the horn of the anvil near the point while the ends are being rounded the piece will be fullered at the middle as shown in 13 *C* and should leave the piece about the correct size, but if it is still too wide through the center, fuller with a $\frac{3}{4}''$ top and bottom fuller to dimensions in Fig. 13 *C*. Punch holes (*a*) and split as shown

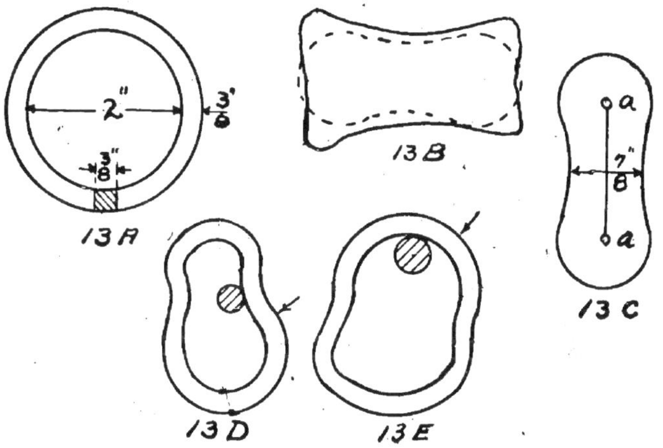

in Fig. 13 *C*. See page 77. Open by driving a punch through the split. Place on the horn and bring to the circular shape as shown in Fig. 13 *D* and 13 *E*.

Example 14. Edge Bend. — Stock, $\frac{1}{2}'' \times 1'' \times 6\frac{1}{2}''$ Norway Iron.

Explanation. The stock is to be bent edgewise to a right angle with a sharp corner on the outer side, which is the essential feature of the exercise. The finished piece must not have any cracks in the corner and must be true to dimensions. Fig. 14 *A*.

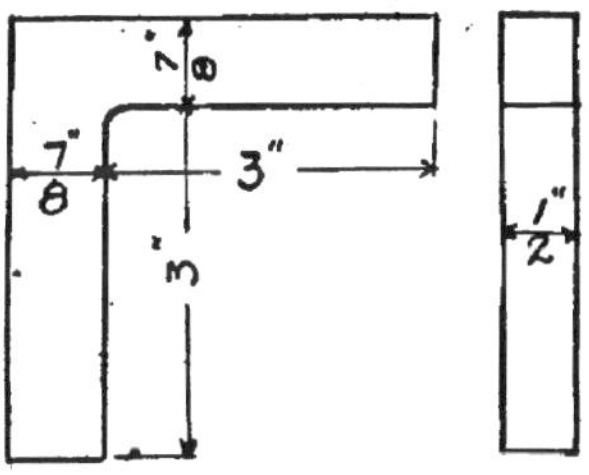

14 A

Operations. All steps are fully explained on pages 72 and 73.

Example 15. Grab Hook. — Stock, Norway Iron $\frac{5}{8}'' \times \frac{5}{8}'' \times 6\frac{1}{2}''$.

Explanation. Stock to be finished to form hook with punched eye. It must be smoothly finished and agree with the dimensions in Fig. 15 *A*.

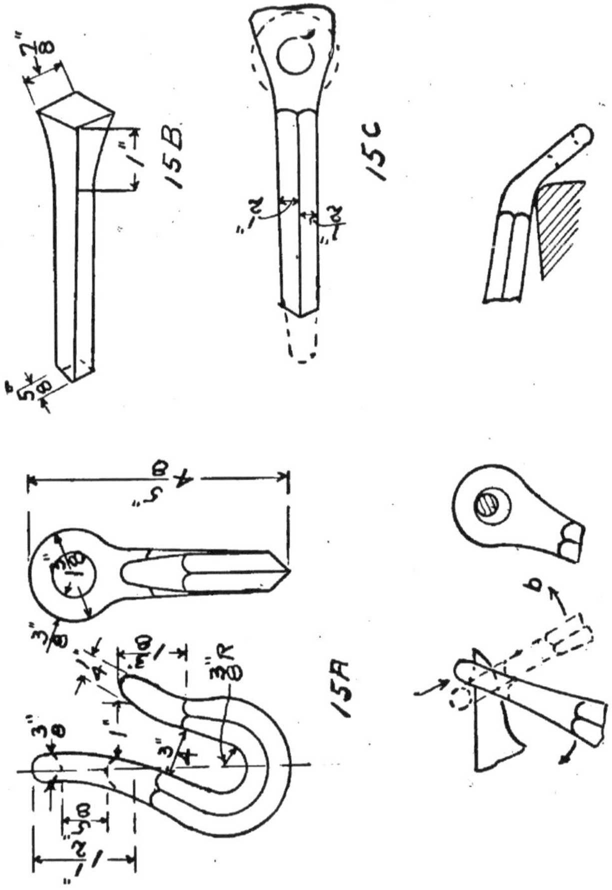

Operation. Upset one end to $\frac{7}{8}''$ square (Fig. 15 *B*). Flatten this upset portion to $\frac{1}{2}''$ thick (Fig. 15 *C*) and round end as indicated by the dotted lines, punch $\frac{1}{2}''$ hole for eye (Fig. 15 *C*). Complete eye by hammering stock around the hole to a circular section over the horn (Fig. 15 *D*), swinging the stock backward and forward and up and down as shown by arrows in Fig. 15 *D*. Draw out the other end to a blunt point as indicated by the dotted lines (Fig. 15 *C*). Bend the eye and point back (Fig. 15 *E*). Bend to the required hook shape. If the sharp edges become flattened in bending they can be brought back to shape on the horn.

Example 16. Nail. — Stock, Norway Iron $\frac{3}{8}''$ round, about $1\frac{1}{2}''$ or $2''$ long.

Explanation. This exercise is to give practice in the use

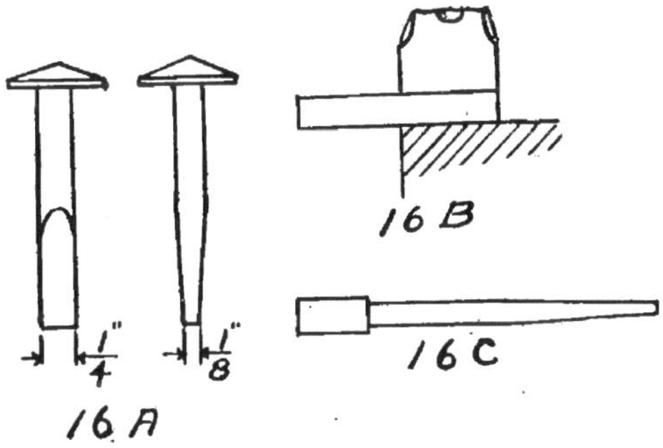

of the heading tool. To get the stem and head concentric make to form shown in Fig. 16 *A*.

Operation. Draw out about $1\frac{1}{4}''$ of the stock with set-hammer as shown in Fig. 16 *B* to the form shown in Fig. 16 *C* and round so it will just pass into the heading tool. Then make head following the directions given on page 88.

Note. The point should be made before the head is formed.

Example 17. Hexagon Head Bolt. — Stock, Mild Steel $\frac{1}{2}''$ round, $6\frac{1}{4}''$ long.

Explanation. The finished piece shown by Fig. 17 *A* must be to size and dimensions, the stem must be straight and at right angles to the head and concentric with it. If the proper amount of stock is not secured in the head, the proper thickness should be maintained at the expense of the other dimensions.

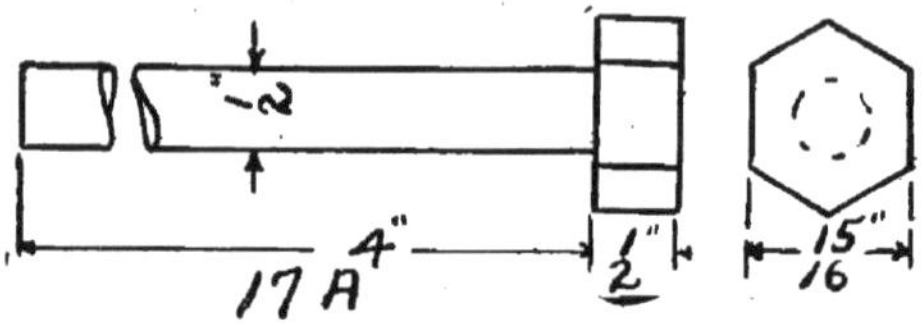

Operation. Follow the directions on page 88 to make the head and on pages 88 and 89 for shaping it.

Example 18. Square Nut. — Stock, Norway Iron $\frac{3}{8}''$ $\times$ 1″.

Explanation. The square nut shown by Fig. 18 *A* is to be made. In case there is too much stock, obtain the proper thickness at the expense of the other dimensions. Adjacent sides must be at right angles to each other and at right angles to the faces.

Operation. Punch $\frac{3}{8}''$ hole in the exact center of the piece. Place on mandrel (Fig. 18 *B*) and with heavy

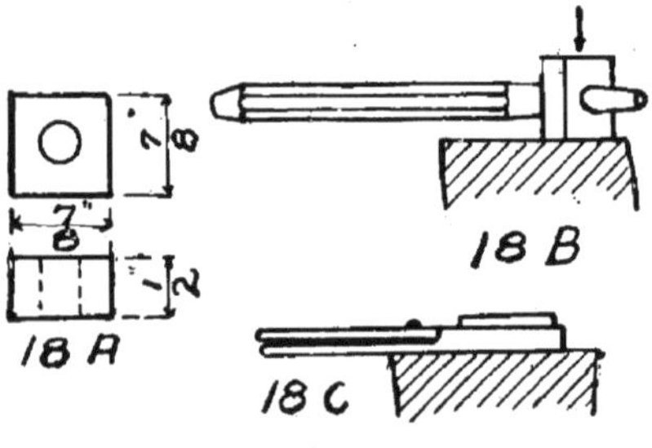

blows bring to size and shape. To true the faces hold in narrow tongs (Fig. 18 *C*).

Example 19. Hexagonal Nut. — Stock, Norway Iron $\frac{3}{8}''$ $\times$ 1″ any convenient length.

Explanation. The nut is to be finished to size and shape shown in Fig. 19 *A*. Faces to be at right angles to the sides, and the sides to form a perfect hexagon.

Operation. Cut the stock as shown by Fig. 19 *B*, bend as shown by Fig. 19 *C* while quite hot, and strike a heavy blow to produce the form shown by Fig. 19 *D*. Bend as shown by Fig. 19 *E* and strike to produce the form Fig. 19 *F*. Punch and break from stock and finish on mandrel.

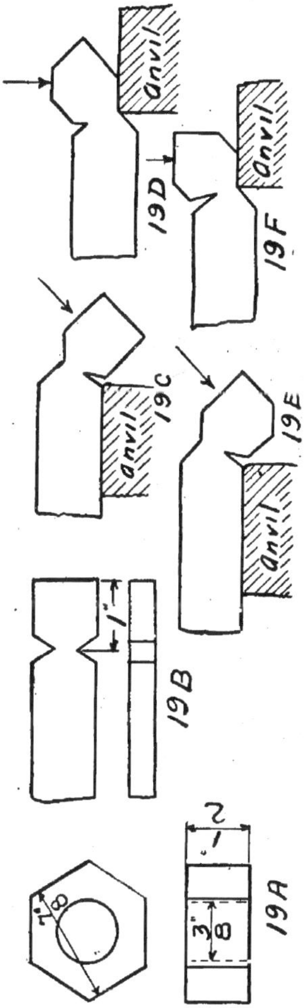

Example 20. Ice Tongs. — Stock, Norway Iron $\frac{5}{8}'' \times \frac{5}{8}'' \times 14''$ (2 pieces).

Explanation. The stock is to be drawn with sledge (or power hammer) to approximately the size shown in Fig. 20 *B* and finished to exact size with hammer, flatter and swage. The handle and blade should be bent to shape, the hole punched and the parts riveted together to form a well-shaped strong pair of ice tongs.

Operation. Draw out enough of one end to $\frac{7}{16}''$ diameter which will stretch to $10\frac{3}{4}''$ long. Smooth with swage. Draw out the balance of the stock to dimensions in 20 *B* and smooth with flatter. Bend handle as shown in 20 *C*, shape blades, punch eye and rivet and case-harden the points.

Example 21. Welding. — Stock, Norway Iron $\frac{3}{8}''$ × $1''$ × $24''$ long.

Explanation. The stock is folded to make three thicknesses and welded to a solid piece and brought

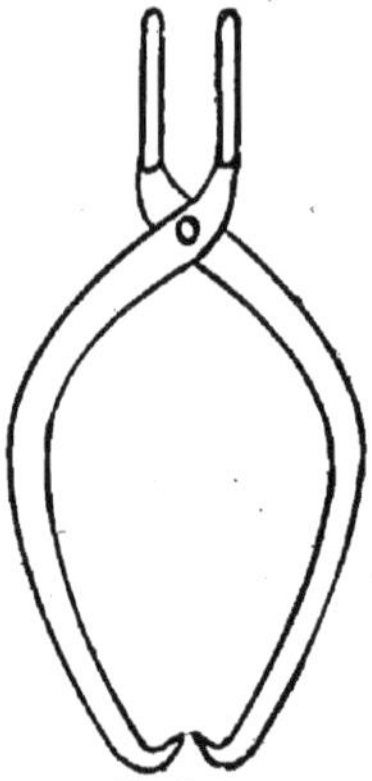

20 A

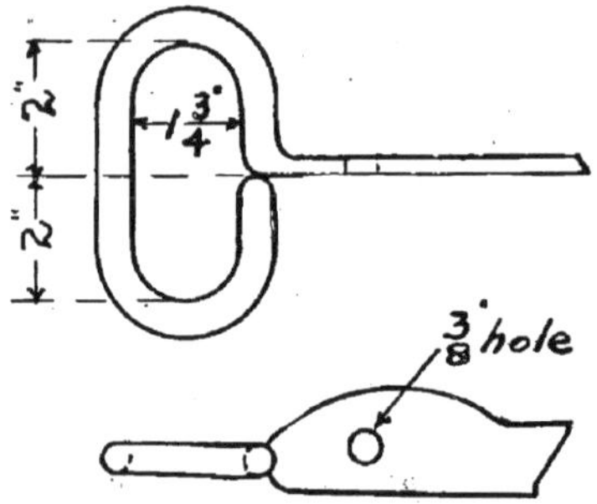

20 C

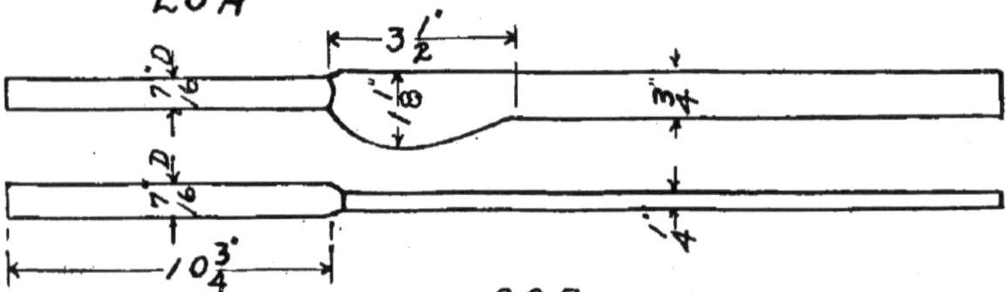

20 B

down to $\frac{3}{4}''$ square section. To test the weld Fig. 21 *A* is made after cutting off the imperfect ends. (The rest of the bar can be used for Example 26.)

Operation. Lay off as shown in 21 *B*, fold as in 21 *C*, and weld. Bring down to $\frac{3}{4}''$ square section. Cut off imperfect ends, fuller with top and bottom fullers $1\frac{1}{4}''$ from end and draw down to $\frac{1}{2}''$ square. (See 21 *A*.)

Lay off center portion $1\frac{1}{2}''$ and fuller all around the stock. Draw down to $\frac{1}{2}''$ round and then form hexagon. Cut off excess stock so as to have the piece the correct length.

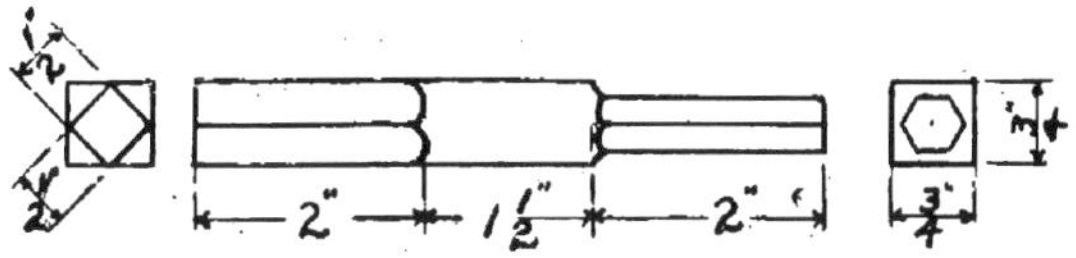

21 A

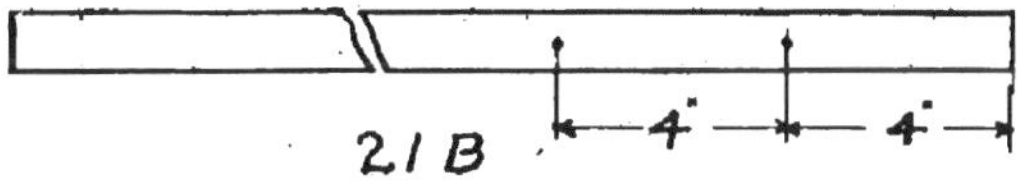

21 B

21 C

Example 22. Chain Link. — Stock, Norway Iron $\frac{3}{8}''$ diameter, $6\frac{1}{2}''$ long.

Explanation. Three links of a chain are to be made for Example 22 but before handing in for credit the three exercises following are made and joined to them by additional links.

Operation. Fully explained on page 95.

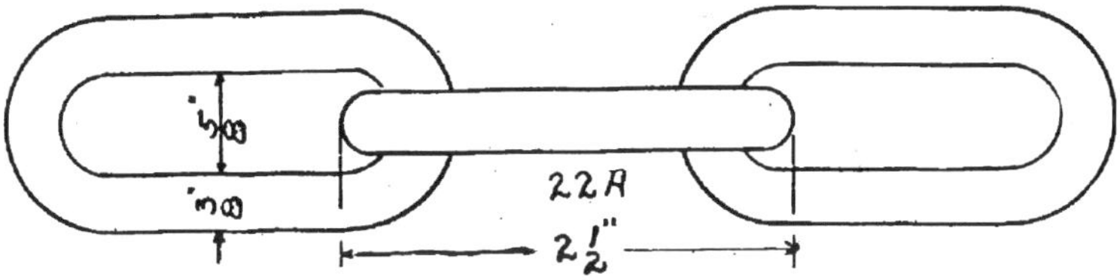

Example 23. Ring. — Stock, Norway Iron $\frac{1}{2}''$ diameter, $9\frac{1}{2}''$ long.

Explanation. A ring is to be made, trued up and joined to Example 22 with an additional link of dimensions of 22 *A*.

Operation. Proceed exactly as when working the link. After welding make as nearly circular as possible on the horn of the anvil and true on the large mandrel. Join to Example 22 with additional link.

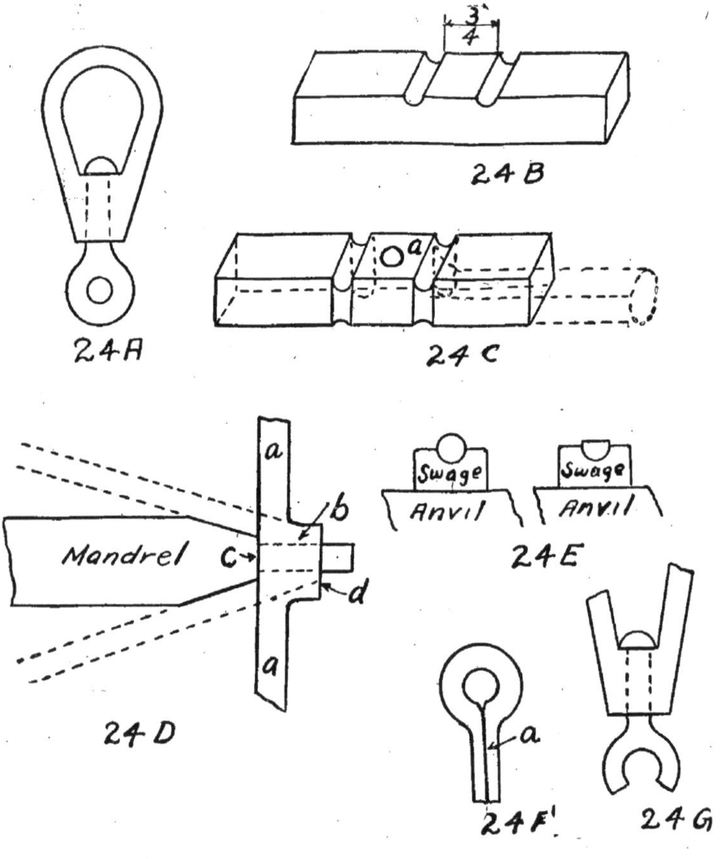

Example 24. Swivel. — Stock, Norway Iron $\frac{3}{4}'' \times \frac{3}{4}'' \times 3\frac{1}{2}''$ and $\frac{3}{8}''$ round $5''$ long.

Explanation. A swivel is to be made (Fig. 24 *A*) and added to the Example 22 with an additional link.

Operation. Center punch at the exact center of the stock and fuller on one side $\frac{3}{8}''$ deep and so the part remaining between the fuller marks will measure a full $\frac{3}{4}''$ square (Fig. 24 *B*). With top and bottom fullers, fuller $\frac{3}{16}''$ deep on the adjacent sides, the same distance from the center. Draw out the ends to $\frac{3}{8}''$ round. Drill hole (*a*) (Fig. 24 *C*). Heat white hot and place on special mandrel (Fig. 24 *D*), and bend the rounded parts

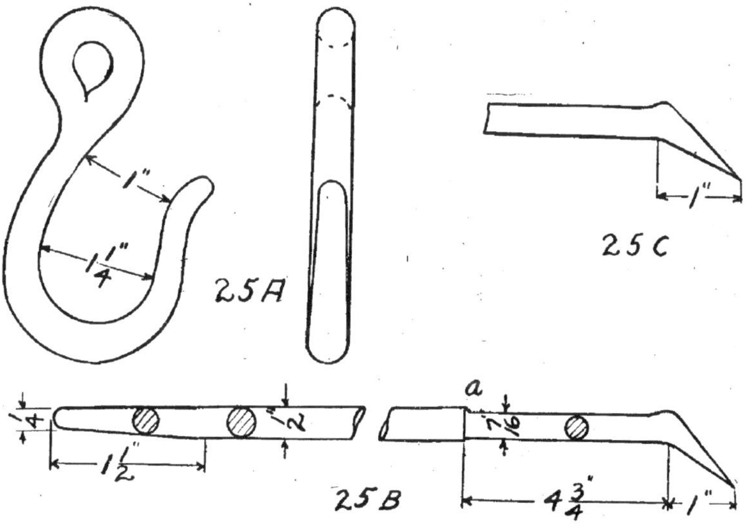

(*a*) as indicated by the dotted lines, and round the portion (*b*) by hammering the corners. Smooth faces (*c*) and (*d*) with a file.

To Make the Eye. Place the $\frac{3}{8}''$ round stock in a swage and hammer each end to half round until drawn out $1\frac{1}{2}''$. (Fig. 24 *E*). Bend as shown by Fig. 24 *F*. Take welding heat on part (*a*) (Fig. 24 *F*) and weld, striking close to the eye first. Draw to $\frac{3}{8}''$ round. Cut off end (*a*) so that about $\frac{1}{4}''$ will project through the swivel and rivet. When riveting hold the eye in the vise. Turn the swivel to keep it from being riveted too

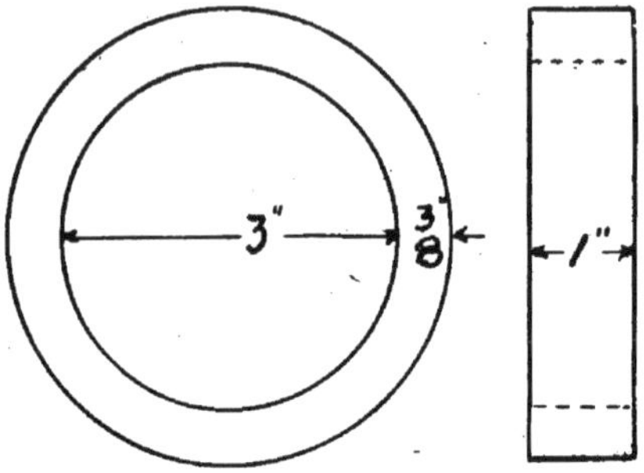

26 A

tightly. Cut off the ends (*a*) (Fig. 24 *D*), so they will measure $3\frac{1}{2}''$ from face (*d*) and scarf and weld like a chain link. Join to chain with an additional link.

Example 25. Hook (Welded Eye). — Stock, Norway Iron $\frac{1}{2}''$ diameter, $9\frac{1}{2}''$ long.

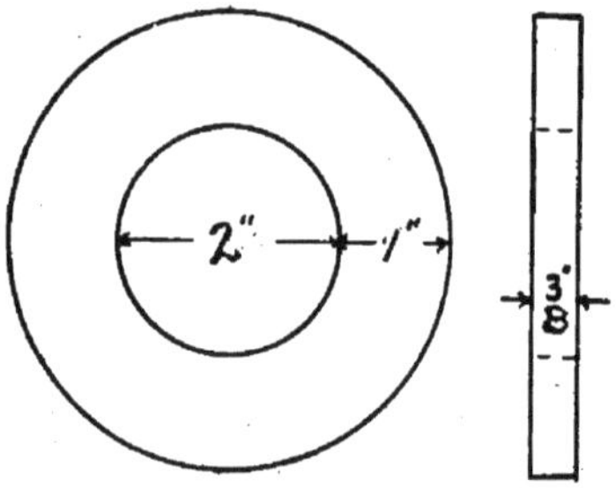

27 A

Explanation. A hook (Fig. 25 *A*) with a welded eye is to be made and joined to the swivel with a chain link.

Operation. Draw down one end to $\frac{7}{16}''$ round and $4\frac{3}{4}''$ long. Draw out the other end to a blunt point (Fig. 25 *B*). Scarf the $\frac{7}{16}''$ end (Fig. 25 *C*). Bend the stock at (*a*) (Fig. 25 *B*) and form eye as explained on page 71. Take a welding heat where the end joins the stem, care being taken not to burn the eye or stem, and weld. Shape the hook to form Fig. 25 *A* and join to the swivel with link.

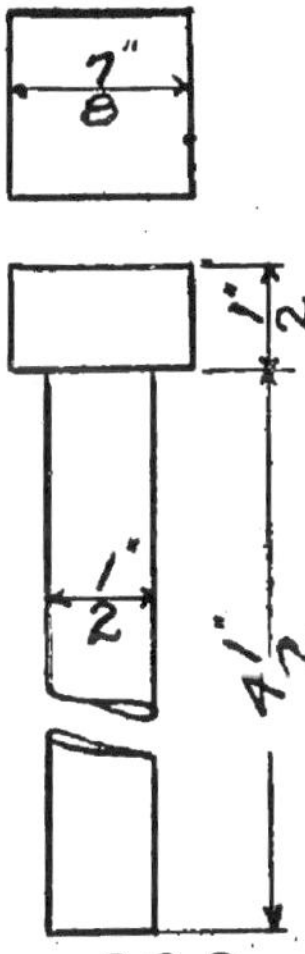

28A

Example 26. Collar. — Norway Iron, $\frac{3}{8}'' \times 1'' \times 10''$.

Explanation. The stock is to be welded to form a collar of dimensions given in Fig. 26 A.

Operation. Is fully explained on page 97.

Example 27. Washer. — Stock, Norway Iron $\frac{3}{8}'' \times 1'' \times 10''$.

Explanation. The stock is to be made into a washer with perfect weld, dimensions given in Fig. 27 *A*.

Operation. Is fully explained on page 98.

Example 28. Bolt (Welded Head). — Stock, Norway Iron, $\frac{1}{2}''$ diameter, 5″ long; and $\frac{1}{2}''$ round, any convenient length.

Explanation. A bolt (Fig. 28 *A*) is to be made by welding a head to the stem.

Operation. Is fully explained on page 99.

Example 29. Two Piece Weld. — Stock, Norway Iron 2 pieces $\frac{1}{2}''$ diameter, 6″ long.

Explanation. The two pieces are to be scarfed and welded so as to form a single straight piece of uniform section throughout.

Operation of welding is fully explained on page 101. The piece should be finished in the swage.

Example 30. Angle Weld. — Stock, Norway Iron one piece $\frac{1}{4}'' \times 1'' \times 3\frac{3}{4}''$, one piece $\frac{1}{4}'' \times 1'' \times 4\frac{3}{4}''$.

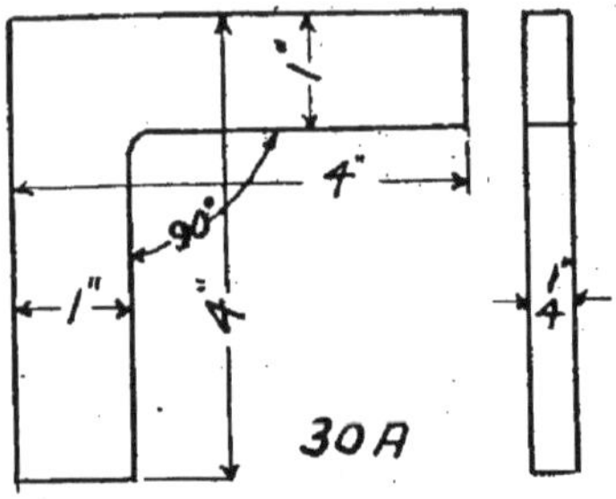

30 A

Explanation. This exercise gives practice in welding two pieces at right angles to each other. The weld must be sound and must agree with Fig. 30 *A*.

Operation of welding is fully explained on page 103. Finish all over with flatter.

Example 31. Tee-Weld. — Stock, Norway Iron, one piece $\frac{1}{4}'' \times 1'' \times 5''$ and one $\frac{1}{4}'' \times 1'' \times 4''$.

Explanation. This exercise while similar to Ex. 30 is more difficult to weld. The finished Tee should agree with Fig. 31 *A*, in form and dimensions.

Operation is explained on page 104.

Example 32. Blacksmith's Tongs. — Stock, Mild Steel two pieces $\frac{5}{8}'' \times \frac{5}{8}'' \times 13''$.

Explanation. Each piece is to be finished to dimensions (Fig. 32 *A*) and these pieces riveted together to form a pair of tongs.

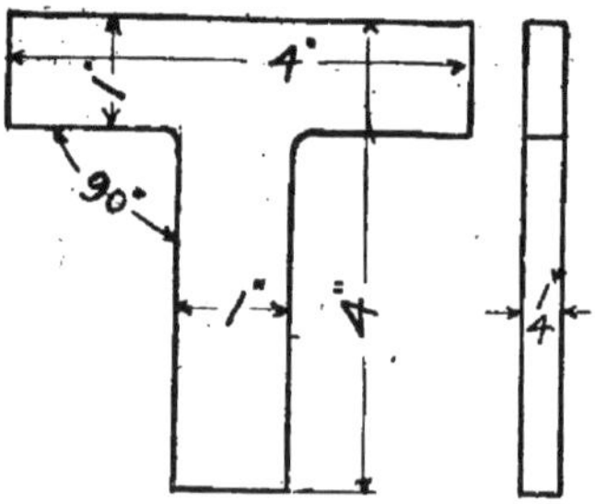

31 A

Operation. Upset one end (Fig. 32 *B*) and fuller (Fig. 32 *C*). Draw out jaw with sledge (Fig. 32 *D*). This will leave the piece as shown by Fig. 32 *E*. Fuller again as shown by 32 F, and the dotted lines (Fig. 32 *E*). Draw out the handle, with sledge (or power hammer), (Fig. 32 *G*) to shape shown by Fig 32 *H*. Punch hole for $\frac{3}{8}''$ rivet as shown at (*b*)

(Fig. 32 *H*). Fuller as shown at (*a*) (Fig. 32 *H*). Finish by drawing the handle to dimensions in Fig. 32 *A* and rivet the two pieces together.

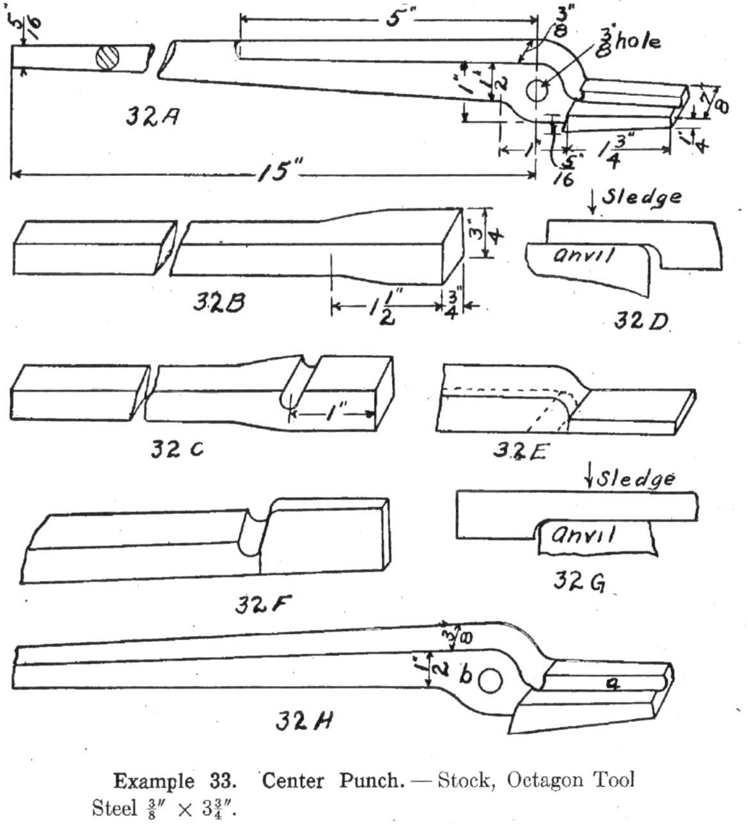

Example 33. Center Punch. — Stock, Octagon Tool Steel $\frac{3}{8}'' \times 3\frac{3}{4}''$.

Operation. Shape end (*b*) (Fig. 33 *A*) to dimensions first, then end (*a*) grind to a point and harden and temper,[1] as explained on page 142.

Caution. In working tool steel great care must be observed not to get it too hot or to hammer it too cold (See directions in Chapter XII).

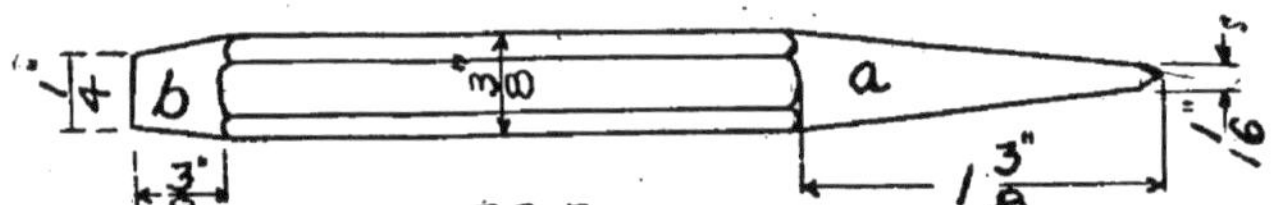

33A

Example 34. Cold Chisel. — Stock, Octagon Tool Steel $\frac{5}{8}'' \times 6''$.

Operation. Shape head end as shown in Fig. 34 *A*. Lay off $2\frac{1}{4}''$ from the other end and shape the blade to dimensions (Fig. 34 *A*). In drawing the blade, lay the stock so that one of its faces will lie flat on the anvil and strike fairly on the upper face. Turn the work occasionally while drawing to keep it straight. Harden and temper as described on page 142.

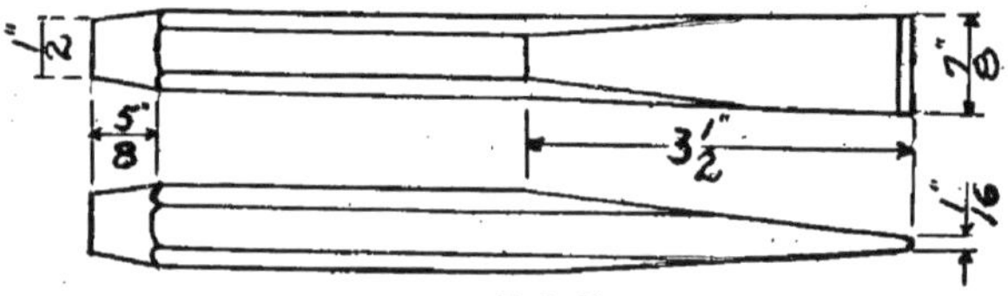

34A

Example 35. Cape Chisel. — Stock, Octagon Tool Steel $\frac{5}{8}'' \times 6''$.

Operation. Shape end (*b*) to dimensions (Fig. 35 *A*). Lay off $1\frac{1}{2}''$ from the other end and shape the blade.

[1] For proper temperature or color to draw temper to, see Appendix, Table I.

In forming the blade fuller to the lines (*a–a*) Fig. 35 *B*. Care must be taken to have the notches the same depth,

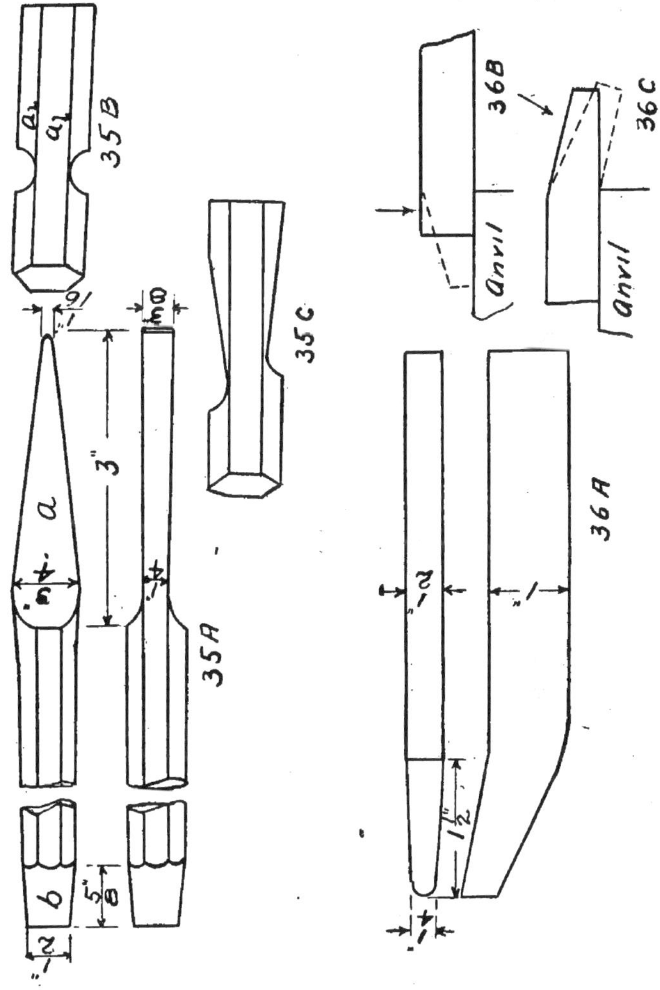

and width, and exactly opposite each other, and not as shown in Fig. 35 *C*. The sides are now drawn out by using a hand-hammer, flatter or sledge. Harden and temper like the cold-chisel.

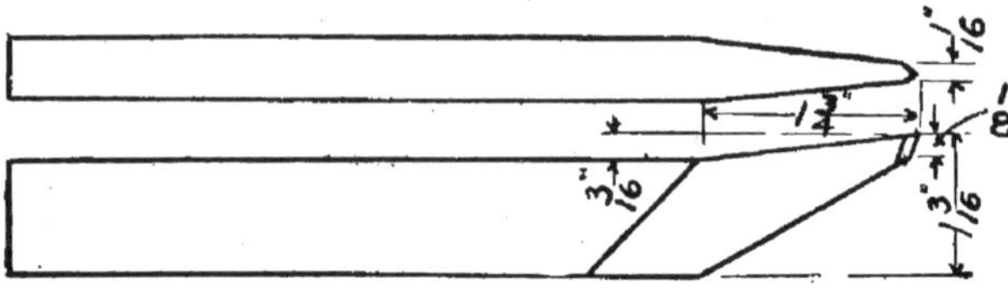

37A

Example 36. Round Nose Tool. — Stock, Tool Steel $\frac{1}{2}'' \times 1'' \times 5''$.

Operation. Hold the stock over the anvil as shown at 36 *B*, $\frac{3}{4}''$ from one end and sledge as indicated by the arrow to the shape shown by the dotted lines. Hold over the edge of the anvil (Fig. 36 *C*)

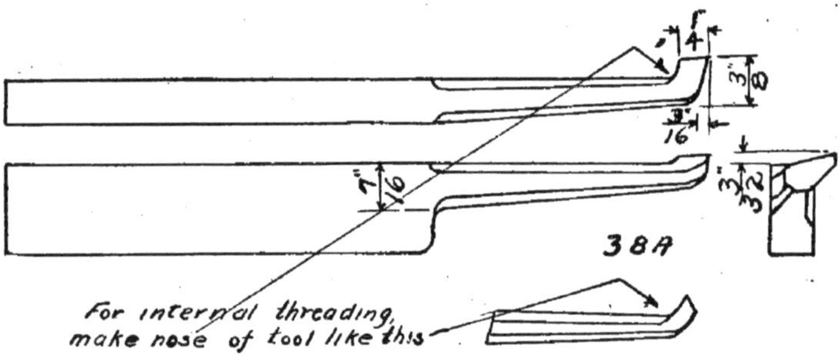

38A

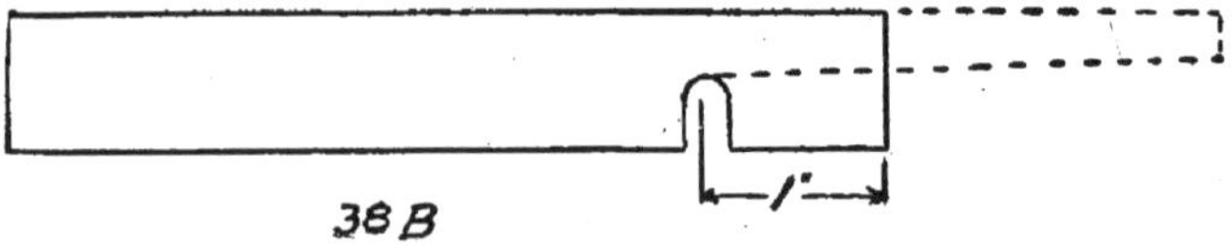

38B

and hit as indicated by the arrow to the position shown by the dotted lines. Taper the sides to the dimensions shown in Fig. 36 *A* and harden and temper as explained on page 143.

Example 37. Thread Tool. — Stock, Tool Steel $\frac{1}{2}''$ × 1″ × 6″.

Operation. Same steps as above for Round Nose Tool, but shape to the dimensions of Fig. 37 *A*. Harden and temper as for round nose.

Example 38. Boring Tool. — Stock, Tool Steel $\frac{1}{2}''$ × 1″ × 5″.

Operation Fuller one edge about half through the stock (Fig. 38 *B*). Draw out end as shown by dotted

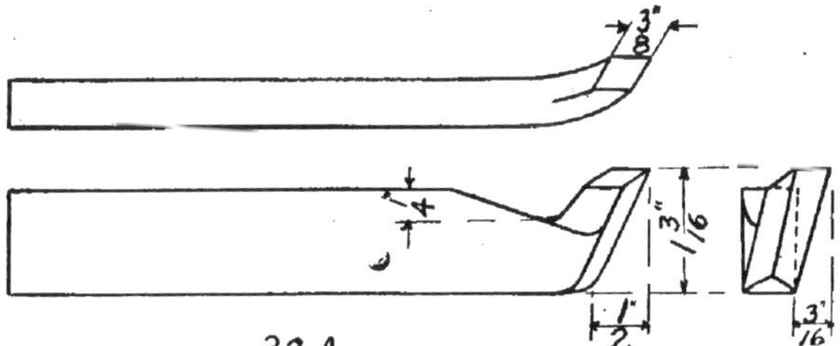

39 A

39 B

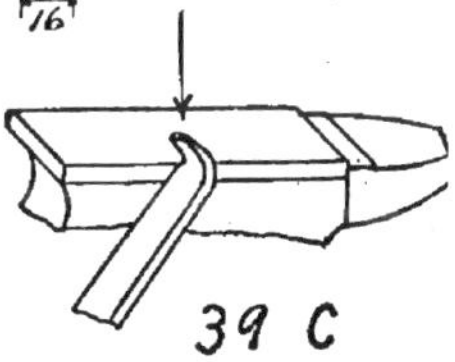

39 C

lines. Bend cutting end and finish to dimensions (Fig. 38 *A*). Harden and temper as tools above.

Note. For some work larger or smaller necks will be needed so it is well to forge tools to different dimensions.

Example 39. Diamond Point. — Stock, Tool Steel $\frac{1}{2}''$ × 1″ × 6″.

Operation. Fuller about $\frac{3}{8}''$ deep, $\frac{1}{2}''$ from one end (Fig. 39 *B*). Sledge to shape shown by Fig. 39 *A*, holding as indicated by Fig. 39 *C*. Harden and temper as directed on page 144.

Example 40. Parting Tool. — Stock, Tool Steel $\frac{1}{2}''$ × 1″ × 6″.

Operation. Fuller half through on one side $\frac{3}{4}''$ from one end (Fig. 40 *B*). Draw out to dimensions shown by Fig. 40 *A* as indicated by Fig. 40 *C*. Harden and temper like diamond point.

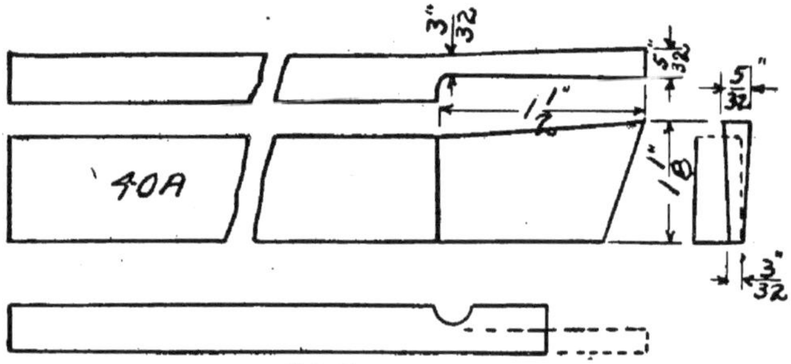

40B

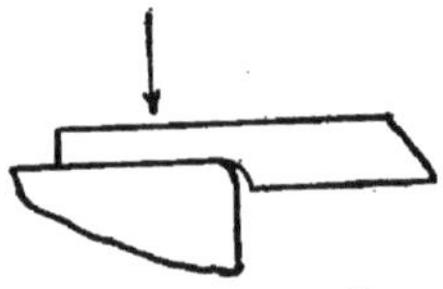

40C.

Example 41. Side Tool. — Stock, $\frac{1}{2}''$ × 1″ × 6″.

Operation. Bevel one edge (Fig. 41 *B*) and fuller (Fig. 41 *C*). With a flatter draw down the portion between the fuller mark and the end, keeping the same slant as the fuller mark which will bring it to the shape (Fig. 41 *D*). The edge *A–B* is made thinner than *C–D* as shown in the section Fig. 41 *A*. Place on the anvil and shape to dimensions Fig. 41 *E*. After all parts are forged to the required dimensions (Fig. 41 *A*) the edge

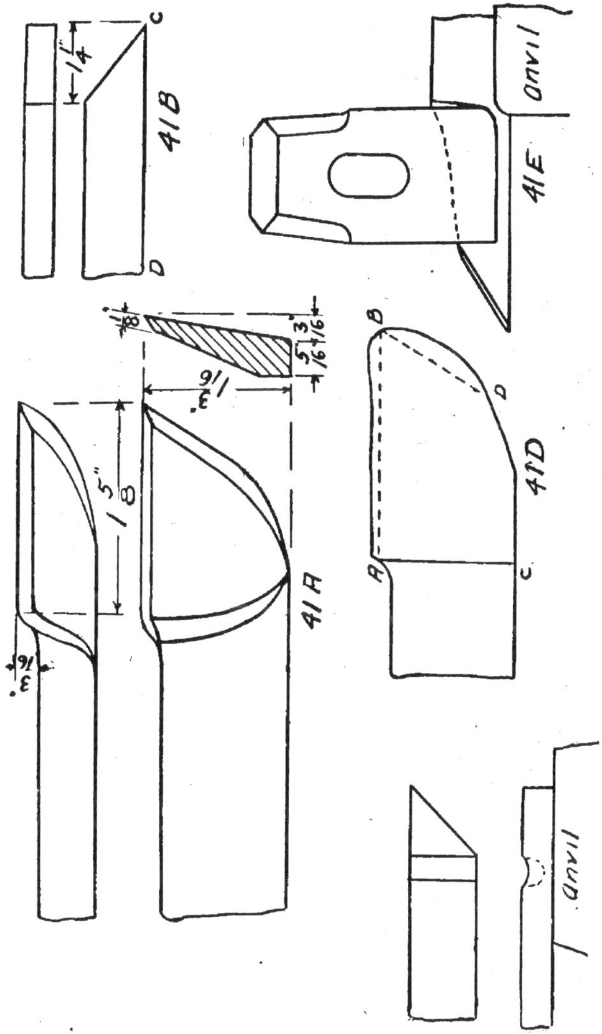
41A
41B
41C
41D
41E
anvil
anvil
A
B
C
D

A–B (Fig. 41 *D*) is offset as shown in the plan Fig. 41 *A*, with the set-hammer as shown in Fig. 41 *E*. Harden and temper as described on page 144.

Example 42. Nippers. — Stock, Tool Steel two pieces ½″ square, 7″ long.

Operation. Follow the general directions in Example 32

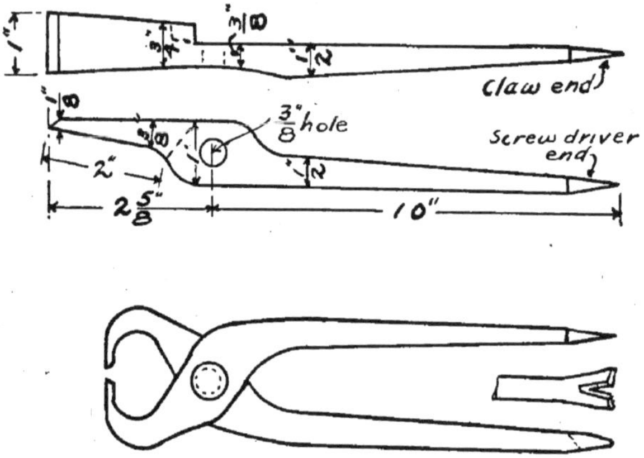

42 A.

and forge to dimensions shown in Fig. 42 *A*. Harden and temper cutting edges only.

Suggestions for Other Exercises. Screw driver, lathe, dog, horseshoe, oar-locks for boat, clevis, wood chisel, gouges, and garden rakes.

TABLE I.—TEMPERATURE AND COLOR TO WHICH VARIOUS TOOLS SHOULD BE HEATED WHEN DRAWING THE TEMPER

Tools	Tempering colors	Temperatures Fahr.
Scrapers, burnisher, hammer faces, reamers, small tools, paper cutters, lathe and planer tools.	light straw	430
Lathe and planer tools, hand tools, milling cutters, reamers, taps, boring bar cutters, embossing dies, and razors.	medium straw	450
Drills, dies, chuck jaws, dead centers, mandrels, drifts, bending dies, and leather cutting dies.	dark straw	470
Small drills, rock drills, circular saws (for metal), drop dies, and wood chisels.	brown	500
Cold chisels (for steel), center punches, scratch alls, ratchet drills, wire cutters, shear blades, cams, vise jaws, screw drivers, axes, wood bits, needles, and press dies.	purple	530
Cold chisels (for cast iron).	dark purple	550
Springs and wood saws.	dark blue	600
Light springs and blacksmith's punches.	light blue	630

TABLE II.[1]—COLOR OF IRON AT VARIOUS TEMPERATURES

Color	Temperature Fahr.
Dark blood red, black red	990
Dark red, blood red, low red	1050
Dark cherry red	1175
Medium cherry red	1250
Cherry, full red	1375
Light cherry, bright cherry, scaling heat,[2] light red	1550
Salmon, orange, free scaling heat	1650
Light salmon, light orange	1725
Yellow	1825
Light yellow	1975
White	2200

[1] Taken by permission from Taylor and White's paper *Trans. Am. Soc. of Mech. Engs.*, Vol. XXI.

[2] Heat at which scale forms and adheres, i.e., does not fall away from the piece when allowed to cool in air.

TABLE III. — POWER AND TIME REQUIRED FOR ELECTRIC WELDING BY THE THOMPSON PROCESS

Area of weld in sq. in.	Watts in primary of welders	Time in seconds	H.P.	Foot pounds
	8,550	33	14.4	260,000
½	16,700	45	28.0	692,000
1	23,500	55	39.4	1,191,000
1½	29,000	65	48.6	1,738,000
2	34,000	70	57.0	2,194,000
2½	39,000	78	65.4	2,804,000
3	44,000	85	73.7	3,447,000
3½	50,000	90	83.8	4,148,000

TABLE IV. — SPEED OF WELDING AND GAS CONSUMPTION FOR OXY-ACETYLENE WELDING

Thickness of plates in inches	Consumption of acetylene cu. ft.	Consumption of oxygen cu. ft.	Speed of work in foot run of weld per hour
0.0394	1.8	2.25	50
0.0591	2.7	3.50	40
0.0787	3.6	4.50	35
0.0984	5.4	7.75	30
0.1181	8.0	10.00	24
0.1575	12.5	15.70	18
0.2204	18.0	22.00	14
0.3071	27.0	33.00	10
0.3582	36.0	44.00	7

Allowance for the Machining of Forgings. — For articles up to 5″ in diameter allow ¼″; from 6″ to 8″ allow ⅜″; 9″ to 10″ allow ½″; and 1 ft. allow 1 in.

Bath for Bluing Steel.

Water	1 gal.
Hyposulphite of soda	2 ounces
Acetate of lead	1 "

Add the hyposulphite of soda and the acetate of lead to the water and heat to boiling. (At first a white precipitate will appear but this soon turns black when near the boiling point and the solution is ready for use.

The steel or iron to be used is cleaned of grease and dipped into the bath until the proper color appears. At first there will be a golden color that rapidly changes to a red and finally a blue. This takes but a few minutes so the piece must be carefully watched and removed the instant the piece is at the proper color, and rinsed and dried. If the pieces are first coppered the colors will be much more brilliant. Brass articles can also be given a very pretty coloring by this bath.

General Tools Required for a Class of 12 Pupils.

3 8-lb. sledges
2 cutters (hot)
2 cutters (cold)
2 top and bottom fullers $\frac{3}{8}''$
2 " " " " $\frac{1}{2}''$
2 " " " swages $\frac{3}{8}''$
2 " " " " $\frac{1}{2}''$

INDEX

Spotted something that isn't quite right?

Let us know and we'll fix it for future reprints. And as a thank you for your diligence, we'll send you a free Sothis book of your choice.

publisher@sothis.press

This page intentionally left blank